A Change in the Weather

Climate and Culture in Australia

EDITORS
Tim Sherratt, Tom Griffiths and Libby Robin

NATIONAL MUSEUM OF AUSTRALIA PRESS

First published in 2005 by
National Museum of Australia Press
GPO Box 1901
Canberra ACT 2601
Phone: +61 2 6208 5000
Fax: +61 2 6208 5145
Email: information@nma.gov.au
Website: www.nma.gov.au

Designed and produced by
Halstead Press Publishers
300/3 Smail Street
Broadway NSW 2007
Australia

National Library of Australia Cataloguing-in-Publication data:

A change in the weather: climate and culture in Australia.

Includes index.
ISBN 1 876944 28 5.

1. Aboriginal Australians - Attitudes. 2. Australians - Attitudes. 3. Human beings - Effect of climate on - Australia. 4. Australia. - Climate - Social aspects. I. Sherratt, Tim II. Griffiths, Tom, 1957- . III. Robin, Libby, 1956- . IV. National Museum of Australia. V. Title.

304.250994

Designed by Judi Rowe
Printed in China by Everbest Printing
Endpaper music: *Historical Climate Patterns in an Aural Time Series* created by Jennifer Kesteven

For each climate variable (Southern Oscillation Index, rainfall deciles, pressure in different cities) a sound sequence was created by transforming the time series of monthly anomalies to a musical scale. Three-four time was used with each month being represented by a semiquaver (a quarter of a note) and thus each bar represents a year, beginning in 1952. If an anomaly persists for longer than a month then the length of the note is extended (ie not all months are represented by an individual note). Thus the natural variations in the climate variable produce the rhythm of the sound sequence. It is important to note that different speeds at which the sound files are played can give different impressions of the time series.

(Four sequences, each over the period 1952–1978).
Front: (1) Southern Oscillation Index (2) Australian rainfall
Back: (3) Atmospheric pressure – Canberra (4) Atmospheric pressure – Darwin

Acknowledgements

We thank the Bureau of Meteorology, the National Museum of Australia and the Australian National University for the practical and intellectual partnership that has made this research and publication possible. We are also grateful for the support of the Australian Research Council through a Linkage Grant for the project, The Human Elements: A Cultural History of Weather in Australia [No. LP0347378].

John Zillman, former Director of the Bureau of Meteorology, and Dawn Casey, former Director of the National Museum of Australia, believed in this project from its beginnings and strongly supported its development. Bill Downey at the Bureau and Mike Smith at the Museum provided valuable support to the editors for this creative institutional collaboration. Neville Nicholls has been a constant inspiration to our work on the intersections of science and culture, meteorology and history.

The History Program of the Research School of Social Sciences at the Australian National University and the Centre for Resource and Environmental Studies, also at the ANU, have fostered this research. Kirsty Douglas and Bernadette Hince have offered essential advice and research support. Jill Nicholls at the Bureau of Meteorology library has also provided valuable and unusual material for the project.

This book began as a National Academies Forum held in September 2002 at the Australian Academy of Science, Canberra. We are grateful to the Australian Academy of the Humanities, the Academy of the Social Sciences in Australia, the Australian Academy of Technological Science and Engineering and the Australian Academy of Science for supporting such an interdisciplinary endeavour. In particular we are grateful to Joy Dudine and Faye Nicholas for their support through the National Academies Forum, as well as the staff at the Australian Academy of Science. Gavan McCarthy and his team at the Australian Science and Technology Heritage Centre (Austehc) have pioneered the collection and management of the archives of Australian science, and their work with the Bureau of Meteorology helped create the momentum for this book.

We appreciate the personal and professional support given by Matthew Richardson of Halstead Press, who attended the National Academies Forum, and guided the publication process with the assistance of Tanvi Mehta. Karen Leary, Thérèse Weber and Denis French of the Publications section at the National Museum of Australia enthusiastically contributed their expertise, as did George Serras, photographer at the Museum. We are pleased that this book features items from the National Historical Collection of the National Museum of Australia, and thereby contributes to the Museum's mission to tell environmental history through objects.

We especially thank the contributors to this volume for their friendship, goodwill and vigorous intellectual engagement.

Tim Sherratt, Tom Griffiths and Libby Robin
Australian National University
Canberra 2004

Contents

The Big Dry by Edd Aragon

published in the *Sydney Morning Herald*, 1 October 2003

NATIONAL MUSEUM OF AUSTRALIA

Elements and Outlooks

With a few strokes of the pen, cartoonists dissect the most complex events, reflecting on the big issues of the day. Each cartoon provides a snapshot of our times, leaving a small time capsule of events and how they were understood. Edd Aragon is an editorial cartoonist and illustrator for the *Sydney Morning Herald*. His cartoon, *The Big Dry*, was an entry in the 2003 National Museum of Australia Political Cartooning Competition. Drought, water shortages and the imposition of restrictions on household watering are ever-present issues in most of Australia's cities. Aragon's deeply symbolic and pessimistic cartoon depicts a harsh industrial backdrop, a parched landscape and the last remaining moisture being sucked from the exposed bones of the country.

1

Human elements

Tim Sherratt

'I say emphatically that the climate has changed', Henry Hodgson told the *Argus* in 1928. The experience of seventy-eight years brooked no denial; summers were milder, and thunderstorms were fewer. 'It is no use telling me that weather bureau statistics do not bear this out', he added defiantly. 'You can do anything with statistics, but no statistics will convince me that the climate has not changed radically.'[1]

It's hard not to have some sympathy for Mr Hodgson, for even as we express our concerns about global warming and educate ourselves about the characteristics of Australia's variable climate, there remains a nagging feeling that somehow he was right. Think back to the boiling-hot Christmases of your youth, to those long weeks spent at the beach, and answer honestly — do you remember summer as being hotter?

If you do, I have some bad news for you — it means you're getting old and slow. The assistant director of the Bureau of Meteorology, W. S. Watt, unsurprisingly rejected Mr Hodgson's assessment. It was largely a trick of memory, Watt suggested, quoting W. I. Milham's judgement that such misconceptions were due in part 'to the fact that the attitude towards life, the amount of energy, and the daily occupations and responsibilities of old people are different from what they were when they were younger'.[2] The heat has drained from us, it seems; our thunderstorms have all dried up.

Exchanges such as these occurred regularly across the twentieth century. Bouts of unusually hot or cold weather brought forth claims that the climate was changing, despite persistent denials from the meteorological experts. 'Almost every person in Melbourne who is not a meteorologist is certain that this is the coldest, wettest, and windiest winter that he remembers', noted the *Argus* in 1935.[3] But even if, as the Bureau maintained, this was more a matter for psychology than meteorology, such climatic contretemps raise interesting questions about the relationship between memory and statistics, between climate and weather.

Climate is what you expect, weather is what you get.[4] A detailed knowledge of climate is derived from averages and statistics, expressed most meaningfully in terms of probability.[5] What is the chance of rain? Accumulated experience guides our expectations. But more than just the available data, perceptions of climate reflect our broader aspirations — our powers of imagination as well as observation. In 1744, even before its boundaries were known, Dr John Campbell argued that the mysterious Terra Australis was located 'Precisely in the richest Climates of the World'. It was, he noted, 'impossible to conceive a Country that promises fairer from its Scituation' [*sic*].[6] Assessments of Australia's climate have veered between optimism and anxiety. To some the climate promised health and vigour, to others it meant only hardship. In the early decades of the twentieth century, as David Walker describes, the supposed dangers of a 'tropical' climate nagged at the ambitions of White Australia.[7] The climate derived from years of observation and abstraction continues to be shaded with climates of opinion and emotion — climates of fear and climates of hope.

Weather, on the other hand, is the shower of rain that catches us unprepared, the gust of wind that whips the fallen leaves about our legs, the unrelenting heat that leaves us drained and irritable. Weather is immediate, local and personal. Weather 'doesn't just happen', argues David Laskin, 'it happens to us'. Weather is 'around us *and* inside us'.[8] Weather is the daily experience of nature, it affects the way we feel and the way we act. In its most dramatic moments — in storms, in heatwaves, in cyclones, droughts, or floods — the weather changes lives. Confronted once again by the violence and cruelty of nature, the nation unites in shock and sympathy, politicians mouth familiar promises, and insurance companies dither. The weather becomes a matter for comment and concern, as each event is catalogued and compared — the biggest in fifty years, the worst for a generation, perhaps even a turning point in the nation's progress. Such events provide markers in the histories of individuals, communities and nations. They serve as points of reference and meaning — they stick in the memory.

Herein may lie an explanation of Mr Hodgson's (and our own) delusion. In 1950, a long wet and cold spell led many Sydneysiders to conclude that the climate was changing for the worse. John Hogan, deputy director of the Bureau of Meteorology, ventured once more to dispel such fears, and offered a useful simile. 'Old people who complain about the changing climate, remember only the peak periods', he explained. 'Looking back over their lives, these periods of exceptional weather merge together like the telephone posts down a long road.' As a result, 'outstanding events' were mistaken as 'normal'.[9]

So not only are we old and slow, but our memory is buggered as well.

But no matter how complete our recollection, climate and memory will never coincide, simply because memory is not an average of the past. Memory is built upon the particular and exceptional. High points and low points may be blunted over time, but they retain their individual significance. Memory generates meaning, not statistics. In a similar way, climate is not just an accumulation of weather. We experience the two differently, our lives lurching constantly between expectation and event, between the idea of climate and the reality of weather. And yet we experience them together, locked in a process of negotiation and transformation — through the weather we know the climate, through the climate we know the weather.

This book aims to explore the cultural space between weather and climate. Sections alternate between the *elements* of weather — rain, sun and wind — and climatic *outlooks* that survey limits, patterns, connections and change. The divisions are meant to be suggestive, rather than exclusive, shifting perspective from the individual to the global, from the past to the future — reflecting the process by which weather itself is parlayed into climate and back again. Preceding each section a series of interludes, featuring collection items from the National Museum of Australia, offer other ways of connecting to the book's main themes. The interludes illustrate the 'human elements', the ways in which Australians have learned to understand their climate, to live with each change in the weather.

A backdrop to history

'The influence of climate on our history is still largely unstudied', Geoffrey Blainey observed in 1971. There seemed no doubt, however, that such an influence existed. As Blainey noted, the Australian climate had been 'a vital determinant of the prosperity and location of the wool industry, dairy farming, and every rural industry ranging from sugar to wheat'. Climate 'helped to dictate the site and size of cities', while drought had 'shaped some of the most dismal eras of our history'.[10] Climate had suffered the neglect of historians simply because its effects had been taken for granted. It was too obvious.

Like the environment, the landscape, or nature itself, climate is often imagined as the backdrop against which history is played out. Climate provides colour and context, but it also frames the limits of the drama. Upon its stage, history is enacted through scenes of achievement and failure. Individuals and nations battle the cruelties of climate, or are blessed by its generosity. Some seek conquest, proclaiming that the climate itself must submit to civilisation. Others adapt to its challenges, and are themselves changed. Such sagas are familiar in a settler society like Australia, where progress was expected to be won in an ongoing confrontation with nature. Like the frontier itself, the climate has wrought its own sense of opportunity and anxiety.

The sun, for example, could be a dangerous foe, or a symbol of new life. From the broad-brimmed cabbage-tree hat to the familiar, shady verandah, European settlers fashioned means to escape its unfamiliar heat and glare.[11] But the Australian sun also offered a new source of light and energy, an escape from the gloom of Britain. In the early twentieth century, as David Walker describes in Chapter 8, 'climatic patriots' seized upon the sun as a symbol of Australian vigour and promise. Blessed with abundant sunshine, Australia is still portrayed as a nation whose life is lived out of doors — on the beach, on the farm, on the playing field — even as we grapple with the highest rates of skin cancer in the world.[12] Like the early colonists, we have returned to the protection of the broad-brimmed hat. The Prime Minister declares his love of this 'sunburnt country' peering out from beneath the ample shade of his Akubra.

'Droughts and flooding rains' have wrought a similar mix of threat and self imagining. The vagaries of the global marketplace have combined with the irregular cycle of drought and flood to chart the limits of Australia's economic ambitions. But the brutal uncertainty of the country's climatic variations has also provided the background against which the resilient character of its people could be celebrated. C. E. W. Bean, the chronicler of the Anzac legend, argued that the qualities of the

Australian fighting man were forged in the battle against the elements. 'The Australian is always fighting something', he argued. 'In the bush it is drought, fires, unbroken horses and cattle.'[13] In times of drought we marvel still at the heroic forbearance of the typical rural 'battler'.[14]

The wind, as Tom Griffiths reminds us in Chapter 13, has also borne both health and disquiet — the bracing freshness of a virgin land and an unsettling whiff of alien climes beset by desert or ice. It was also the wind that brought settlers, explorers and invaders, swept along by the 'roaring forties' in a dramatic transition from old world to new. Wind powered the early development of the colonies, providing means of transport and communication without which they could scarcely hope to survive.[15] But as European settlement moved north into lower latitudes prowled by tropical cyclones, white Australians also came to know the wind's destructive power.

The Australian climate has proved both inhospitable and welcoming, rich in its promise and profound in its terrors. It has also been prone to miscalculation, confounding European expectations from the time of the Dutch traders, who misjudged the winds and were blown upon the coast of a seemingly barren land. On the eastern side of the continent, Blainey recounts, Cook and Banks arrived at Botany Bay in the midst of a wet autumn and were impressed by lush meadows and a swiftly flowing stream.[16] Their encouraging report strengthened plans for occupation, but when the First Fleet landed they found a dry and difficult land. They had arrived, as Richard Grove explains in Chapter 11, in the midst of one of the most significant El Niño events in recorded history.

A new climate cannot be mapped and comprehended like a new continent. It can only be known through time, through averages and extremes, through experience and expectation. Over the years we have become familiar with the patterns and trends, we know when the frosts come, when the temperature soars, when the winds are at their strongest. But we have also invested these patterns with meanings that cannot be read from the thermometer or rain gauge. Temperature is cross-indexed with race and health, drought provides an indicator of destiny. Our efforts to understand the climate have merged with the imagining of nationhood, with the construction of an Australian identity. Just as the boundary between society and nature is always on the move, so climate and culture create each other across a shifting, permeable frontier. The backdrop is repainted again and again, as each new actor takes the stage. Climate does not simply shape history, for it can only be known through history.

Meteorological events

In 1788, as a major El Niño event affected weather across the globe, Australia's first white settlers watched, with increasing desperation, as their crops withered and streams dried to a trickle. If all efforts to resupply the colony had failed, if the drought had continued a little longer, whose history might we be writing now?

The influence of climate on history is read not only through patterns and perceptions, but also through events. Droughts, floods and storms intervene in our lives, in ways often dramatic and unexpected. They puncture the confidence of our predictions, temper the arrogance of our plans; they fracture relationships, reinforce communities; they provide moments of collective shock, resolve and recrimination; they divide history into periods both 'before' and 'after'. No doubt most of us could, without too much difficulty, begin to compile our own catalogue of events — Cyclone Tracy perhaps, the dust storm that blanketed Melbourne in 1983, the 1999 Sydney hailstorm, the 2003 drought and bushfires?[17] And yet, although they seem so easy to identify, such events are not always easy to isolate.

In July 1995, the city of Chicago in the United States suffered under record high temperatures. Day after day the heatwave smouldered on, its effects measured, most tragically, in the hundreds of corpses piled up in refrigerated trucks outside the overwhelmed medical examiner's office. Over 700 people are estimated to have died as a result of the heat.[18]

An official investigation described the heatwave as 'a unique meteorological event'. It was a 'freakish disaster', city leaders argued, a brutal whim of nature whose effects could have neither been predicted nor prevented.[19] But heat could not alone explain the dimensions of this tragedy. What seemed to be an act of nature was, Eric Klinenberg suggests, 'a social drama that played out and made visible a series of conditions that are always present but difficult to perceive'.[20] Poverty and isolation left many elderly residents vulnerable. Their experience of the heat could not be understood solely through the temperature on a particular day; it was an experience many decades in the making, brought on by government policy, urban development, and a society that seemed to care little for its weakest and most disadvantaged members. The heatwave was a 'cultural event' whose conditions did not simply 'disappear when the temperatures moderated'.[21] 'We have collectively created the conditions that made it possible for so many Chicago residents to die in the summer of 1995', Klinenberg argues, 'as well as the conditions that make these deaths so easy to overlook and forget'.[22] We are all implicated in such events — the weather does not kill alone.

Meteorological events seem to offer simple lessons in cause and effect. We can track their progress on weather maps and radar screens, we can measure their results with rain gauges and thermometers. Even though a sudden storm might remind us of our vulnerability before nature, it is an event, nonetheless, that comes with its own explanation. The suffering and destruction it wreaks might seem unexpected and unfair, but it is still understandable. Meaning and responsibility are contained within the boundaries of the event. But even as we find comfort in the certainty of markers and maps, our experience of weather refuses to be explained away. As Janet McCalman demonstrates in Chapter 9, heat alone could not account for variations in infant mortality amongst poor immigrant populations in nineteenth-century Melbourne. Women and their babies did not merely suffer the heat of an unfamiliar climate, they suffered the heat of an overcrowded, unventilated slum. They suffered a heat which challenged their understanding of child rearing and health. They suffered a heat whose meaning is to be found both in collective experience and individual life history.

Australia's emergency management authority estimates that in the last hundred years heatwaves 'caused more deaths than any other natural hazard (except disease), yet they remain one of the least-studied and most-underrated' phenomena.[23] As Klinenberg notes, heatwaves do not generate the spectacular images of destruction favoured by the media. When the temperature soars, it is images of crowded beaches we see on the news, not the elderly and the poor dying alone. Unlike floods or fires, heatwaves fail to inspire a sense of community concern. They are 'silent and invisible killers of silenced and invisible people', rarely figuring in our catalogue of national disasters.[24] In 1939, massive bushfires raged across much of south-eastern Australia, culminating in the horror of 'Black Friday'. Seventy-one people lost their lives. Amidst the grief and recrimination, a Royal Commission was called to investigate the causes of the disaster and prevent its recurrence.[25] And yet, in the heatwave that helped fuel the fires, it is estimated that more than 400 people died. These deaths brought no inquiry, no demands for action. The circumstances which claimed their lives were too diffuse to be named, too complex to grab attention. Instead of being remembered as a traumatic moment in the nation's history, the heatwave and its human toll merely blurred into the flow of time.

Tourist poster of the 1930s by Harry Kelly.

TASMANIANA LIBRARY, STATE LIBRARY OF TASMANIA

Meteorological events have beginnings and ends determined not only by our observations, but also by our systems of classification and measurement, by our values and priorities. Of course, such definitions themselves have a history — their own beginnings, their own ends. In 1896, H. C. Russell, the New South Wales government astronomer, noted that the word 'drought' was used differently in Australia from in England. Instead of being defined as 'a period of a few days or weeks in which not a drop of rain falls', a drought in Australia was understood as 'a period of months or years during which little rain falls and the country gets burnt up, grass and water disappear, crops become worthless and sheep and cattle die'.[26] A recent survey catalogued 150 definitions of drought worldwide, reflecting the variety of climatic and cultural conditions in which the definitions were framed.[27] In Australia, as Russell observed, drought has tended to be defined in an agricultural context. With the nation's economic health heavily dependent upon its rural industries, drought has loomed, not just as a climatic oddity, but as a serious threat to progress.

Recent years, however, have brought a further shift in the definition of drought that informs government policy.[28] No longer is drought regarded as a natural disaster, instead it is accepted as an inevitable characteristic of Australia's variable climate. It is a risk that the successful farmer must learn to manage. This seems a rather overdue acceptance of the reality of drought, a sign that we are seeking to work more within the limits of our environment, rather than waging war against it. But what does this mean to those struggling with the immediate effects of drought? Is it any less of a disaster to a farmer forced to shoot his stock, to a community steadily drained of its population, to a woman who watches her lovingly-tended garden wither, or to a young child who has never known the land to be green? Like the suffering of an elderly person in Chicago, or an immigrant mother in North Melbourne, the lived experience of drought is not contained within the event itself.

Daniela Stehlik and her colleagues have tried to develop an understanding of drought as a lived experience by interviewing farm families in New South Wales and Queensland afflicted by drought in the early 1990s.[29] Their study found that the idea that drought was a recurring phenomenon whose effects could be mitigated by careful management was hardly new to farmers. But even if the principles of risk management were scarcely controversial, the implications of the policy shift added considerably to the emotional burden of farm families. A farmer in need was now portrayed as a 'bad manager', proudly independent folk were being lectured on the benefits of self reliance, government policy and the attitudes of urban dwellers seemed to be hardening against the bush. The cumulative effects of political, economic and social change made farm families feel as if they were being abandoned.

Attempting to define drought, one female beef producer commented, 'It is a long period without rain. It is the stress and the hardship. We had a financial drought ... prior to going into the climatic drought'.[30] Climate was only one component in the experience of drought which included short-term market fluctuations, long-term economic adjustment, the steady withdrawal of services from rural areas, and the growing fragmentation and dislocation of rural communities. For these families, at this time, the experience of drought was one of increasing isolation.

Even under a risk management regime, uncertainty remains. The vagaries of the global market can wreak havoc upon the most carefully thought out plans. But there is a more fundamental uncertainty at the heart of the drought experience. One New South Wales sheep and wheat producer remarked in an interview:

> So to try to manage the ups and downs of not knowing are we really in it, are we facing a really severe drought or is [this] just a dry period? I think that is one of the things we have found hard in this. [31]

When do you know a drought has started? Another sheep and wheat producer commented:

> ... the drought just doesn't change [things] overnight does it? It comes on slowly and the country here has changed [even] in the last three weeks. After that green tinge we had from that rain in February, it is all gone now so it is continually changing and you can't really pinpoint a period of time I don't think. It's not like one day it's hot and the next day it's cold. The drought is not like that ... [32]

Droughts are easy to define in retrospect, though it might be noted that the process of drought declaration involves a number of political as well as climatic calculations. But a drought is also made up of a seemingly endless succession of days, each bringing its own weather, each met with a combination of hope, anxiety, resignation and disappointment. Somewhere along the line the disappointments are all gathered up and the word 'drought' is uttered. And yet, even then the days continue, each offering still the prospect of change — perhaps it will rain today? The experience of drought is of a thousand hopes dashed, a teetering balance between expectation and despair. Uncertainty becomes a way of life.

The national outlook

Cyclone Tracy was an event that changed lives. In Chapter 14 Bill Bunbury lets us listen to some of the 'voices on the wind', to the recollections and reflections of those who lived through the terror. The experience of Darwin residents, he suggests, reflects the cyclic movement of the cyclone itself, with the initial swathe of destruction followed by phases of dispersal and renewal. What was left at the end of the cycle was a new city, full of old and difficult memories.

For the local Larrakia people, as Deborah Rose explains in Chapter 3, the cyclone carried a different meaning. Darwin did not suffer through some arbitrary whim of nature, the cyclone was a strike against those who were reluctant to grant land rights to the Larrakia people: 'it came to a particular place — Darwin — because that was where the problems were'. Amongst Aboriginal people in the Kimberley, on the other hand, the story of the cyclone is remembered as 'a warning to keep their knowledge and culture strong'.

Cyclone Tracy had specific, local meanings, but, as Brad West points out, it was also the first cyclone to be 'interpreted as a national event'.[33] While droughts, floods and fires had long found a place in the national mythology, cyclones had previously been accorded only regional significance. The scale of devastation, the involvement of the military, and the dispersal of Darwin residents around the country, all helped to ensure that Cyclone Tracy would be understood differently — as a symbol of the defining battle between nation and nature. Cyclone Tracy joined the landing at Gallipoli as an imagined moment of unity and clarity, calling the supposed characteristics of our national identity to the fore.

Meteorological events can be invested with meanings both personal and political, but they also exist on a geographical scale that positions them locally, regionally and nationally. In his social history of English weather, Vladimir Jankovic describes how meteorology in the seventeenth and eighteenth centuries was focused on the reportage of unusual local phenomena.[34] Descriptions of unexpected or remarkable visitations were reported at scientific gatherings and disseminated through a variety of publications. These eyewitness reports were not merely accounts of the weather, but commentaries on specifically local circumstances infused with ideas about politics, aesthetics and spirituality. By the late eighteenth century, meteorology was beginning to take a more quantitative turn — shifting from colourful descriptions to carefully collected data, from local curiosities to global systems. The 'meaning of locality changed', Jankovic argues, from 'an exclusive end of investigation to a specimen in a larger entity, a point on a grid'.[35]

This change in emphasis coincided with the European settlement of Australia, challenging the fledgling science with an immense task of coordination. Vast networks of volunteer observers were assembled to gather local data, while new technologies such as the telegraph, radio, radar and aircraft were deployed against barriers of distance and isolation.[36] The need for cooperation figured prominently, as meteorologists sought to gather the sprawling continent within the ken of science. By the 1870s, observatories around the country had established an intercolonial framework to aid in the setting of standards and the sharing of information.[37] As Australians began to consider the possible benefits of Federation, meteorology offered an example of the efficiencies to be gained through a coordinated national system. At the 1897 Federal Convention, Sir Joseph Abbott successfully argued that the Commonwealth be granted power over meteorological observations to ensure 'uniformity throughout Australia'. 'If there is anything which ought to be the subject of a Commonwealth law', he continued, 'it is these observations'.[38] The study of meteorology was a national obligation.

This theme was elaborated by the Minister for Home Affairs, Littleton Groom, as he introduced legislation to create a federal meteorological department in 1906. National coordination offered improvements in efficiency and accuracy, but also, he noted, the opportunity to extract a broader understanding from fragmented, local experience. 'Under a Federal system', Groom continued, 'we hope to become thoroughly acquainted with the climatic conditions of the continent'.[39] Knowledge of the Australian climate would develop alongside the fulfilment of Australian nationhood.[40]

But what is an 'Australian climate'? In a continent that ranges from rainforest to desert, from temperate to tropical, we still imagine ourselves as somehow united in an expression of climatological nationalism. Our climate may be something to be celebrated or cursed, but it is still *our* climate — advertised to the world as the symbol of our distinctiveness. Meanwhile, confidence in the idea of a single 'Australian identity' has begun to wane. We have become rightly suspicious of blanket terms such as 'Australian culture', seeking instead a more sensitive approach to issues of class, gender and power. What about climate? What fractures and conflicts are obscured by our continued desire to create a sense of unity from a diverse range of meteorological experiences? Is the sun that warms a homeless man in Hobart the same sun that tans a parade of Gold Coast pleasure seekers? Is the wind that lifts the topsoil in western New South Wales the same wind that flutters the flag atop Parliament House?

There are obvious divisions, of course, as cities snipe and skirmish over the quality of each other's weather. Who has the most 'liveable' climate? Whose is the driest or sunniest? Sydneysiders can barely mention Melbourne without passing judgment on the fickleness of its climate, while Queensland claims the sunshine as its own, extolling an idyllic lifestyle where the weather is always 'perfect'. Within the idea of an Australian climate exist a series of regional stereotypes, deployed in the service of parochial politics and tourism marketing boards.

But there are subtler tensions as well, focused on the continuing significance of locality. Describing the operations of the Weather Bureau in 1922, an article in the *Argus* likened it to 'a huge nervous system, having its centre in Melbourne, with its ramifications stretching to all parts of Australia and contiguous countries'.[41] The organisation prided itself on being the only weather service in the world to encompass an entire continent. However, this mastery of space had its limitations, particularly when it came to delivering useful and timely information to regional areas. At a conference in 1964, the Bureau admitted the need for greater 'regionalisation' of its services.[42] 'The Gippslander feels his needs are not met by a forecast from Melbourne', noted the meteorological officer in charge of the RAAF station at East Sale. 'In hill country such as this', he added, 'the changes in climate from one valley to another are too great to be covered by general forecasts'.[43]

This was not just a question of size or scale, however; it was also a matter of where useful knowledge was generated. W. M. Prescott, a dried fruits and citrus grower from Red Cliffs, told the conference how growers had organised themselves into a 'frost club'.[44] With the assistance of the local weather office, club members closely monitored conditions, sending out an alarm when frost seemed likely. 'Farmers can read the immediate weather signs', grazier A. R. Johnston remarked, 'often with surprising accuracy'.[45] What was needed from the Bureau, he argued, was a complementary picture of what lay beyond the horizon. Rural producers were not merely seeking detailed regional information, they were looking for ways to integrate their local knowledge and experience within the centralised systems of the meteorological experts.

The shift, that Jankovic observes, in what counted as meteorology, did not mean that older forms of observation and reportage simply stopped. Instead, they were gradually squeezed out of the metropolitan literature. The cataloguing of local phenomena and meanings continued, and continues, but is only deemed to add to science if its methods are standardised and its results collated. Folk wisdom is not tabulated, cautionary tales are not compared. In the new meteorological networks that developed in the nineteenth century, knowledge was generated at a distance; meaning and authority flowed from centre to periphery. But we do not experience weather as simply a scientific phenomenon.[46]

The stresses of drought, Daniela Stehlik reports, are often exacerbated by the ill-considered advice of outside experts. Perceived as coming 'from what is effectively a foreign culture', experts and policy makers offer little respect to local experience.[47] It seems they do not come to listen. Scientific expertise is assumed to be concentrated in centres of learning or power. As in the centralised networks of modern meteorology, knowledge is gathered, sorted and tested, before being redistributed as truth. Mere experience is devalued in favour of controlled observation, received truths are displaced by experiment. We accept such changes in the name of progress, while clinging to the belief that there are other ways of knowing — there are truths to be found in place, people and self that cannot be deduced from the canons of science. Into this uneasy truce the weather introduces a daily challenge. Every day we venture out into the world equipped both with forecasts and feelings. To the experts' predictions we bring our own repertoire of signs, memories, and intuitions. Is the breeze warmer than yesterday? Is that rain I can smell? Meteorologists gather data from around the globe, running it through computer models of great complexity. They watch from space as cloud masses dance and twirl across the continent. All this information is collated and processed to tell us what to expect in our own region, city or suburb. But it does not tell us what to feel — at the end of all these transformations and shifts in scale, the weather remains something immediate and personal.

Even our understanding of the global climate is coloured by local concerns. After surveying residents in Newcastle, Harriet Bulkeley concluded that public understanding of climate change could not be measured solely in terms of the spread of scientific information.[48] Just as seventeenth-century observers had framed their reports of meteorological phenomena within a broader social commentary, so inhabitants of Newcastle drew upon 'local knowledges, values and moral responsibilities' in piecing together their understanding of global environmental concerns.[49] Climate change was located within a broader critique of humanity's interaction with nature, but it was also related to the coal-burning furnaces of the local BHP plant. In a similar way, regional variations in the uptake of solar hot water systems cannot be explained by economics alone. As Ian Lowe comments in Chapter 10, the acceptance of such technologies is dependent upon a number of social and political factors. And so even as Queensland proudly celebrates its sunshine, the state lags behind others in utilising this source of renewable energy.

We are frequently reminded that communications, economics, and culture — perhaps even life itself — are becoming increasingly globalised. National boundaries are giving way to the flow of ideas and money. Weather and climate seem to be following the trend: demanding world attention as never before, uniting nations in concern. As Tony McMichael and Clive Hamilton remind us in Chapters 12 and 16, global warming brings a new perspective to Australia's understanding of its climate. Questions of national development have to be reframed within the context of global responsibility. In *Strange Weather*, Andrew Ross argues that the debate about global warming generates 'contradictions in the way people think about the natural world': 'Instead of feeling the weather as we have felt it historically, as part of a shared local, or even national, culture, we are encouraged to think of it globally'.[50] Once a symbol of our distinctiveness, the Australian climate now also reminds us of our obligations to the world.

Even the cycle of drought and flood, so central to Australian experience and identity, is now shared with the world through a massive system of ocean–atmosphere interaction. The farmer in Queensland despairing of drought, feeling isolated from sympathy and support, might yet be linked through El Niño to weather events more than an ocean away. From local to global, from feeling to knowing, the experience of weather presents us with dramatic shifts in scale and meaning. Such shifts connect our choice of clothes for the day with the onrolling of immense oceanic currents. A decision about when to plant our backyard tomatoes might reflect the imagined destiny of the nation, while the opening of the ski season could wait upon our decision about whether to drive or walk to work.

Strange weather

Weather 'is the last thing we leave behind' when we move, David Laskin writes, 'and the first thing we find when we arrive'.[51] As we travel, we carry a sense of 'our' weather amongst the baggage of memory and emotion. The weather we grew up with, the weather we know, provides us with a basis for comparison, encouraging us to declare 'all weather that departs from it strange'.[52]

Newspaper headlines betray our continuing fascination with 'strange weather' or 'freak weather' — weather that is unseasonal, unusual or unexpected. Like seventeenth-century English observers, we greet such visitations with curiosity, and sometimes alarm. Perhaps, like Mr Hodgson, we read such events as evidence that the climate itself is changing; or maybe they warn of the dangers unleashed through human interference with nature. The statistical analysis of climate may often reveal 'freak' occurrences to lie well within the bounds of probability, but their 'strangeness' is also measured against our perceptions of history, memory, place and politics. 'Strange weather' exists in the gap between weather and climate, between hope and fear, between expectation and event.

For Mr Hodgson, it was the present that was strange. Alienated from his imagined past, uncomfortable with the encroachments of age, he looked to the weather as a sign of what had gone wrong — what he had lost. For Cook, Banks, and those who followed, it was the land that was strange. Carrying the weather of home in their memories and expectations, they found it hard to read the southern skies. To European eyes, Australia seemed a place where all things were 'queer and opposite'. The seasons were upside down, of course, but they also behaved less predictably. Rainfall could vary dramatically from year to year, particularly in the arid centre. In Chapter 4, Libby Robin explores the way in which assumptions about seasonality influenced understanding of Australia's desert regions. The experience of the northern hemisphere led naturalists to expect regular, seasonal cycles in the breeding habits of birds. However, the Banded Stilt refused to conform, only breeding when floodwaters create suitable wetlands in the arid zone. Like many other birds and animals, the Banded Stilt had found a way to adapt to Australia's variable climate. Instead of regularity or seasonality, there was opportunism.

As early as the 1850s, Neville Nicholls notes in Chapter 2, William Stanley Jevons had characterised the Australian climate as 'one of irregular rains'. But the significance of this observation has been difficult to grasp. While periodic droughts were reluctantly accepted as features of Australian life, the idea that variability of this sort was not an aberration, but a defining characteristic of the climate, seemed to conflict with ideas of natural order and balance. This feeling of unease can be observed in the reception of the well-known 'Goyder's Line'. In Chapter 5, Janis Sheldrick describes how George Goyder, Surveyor-General of South Australia, sought to map the limits of reliable rainfall, which he recognised as crucial to the colony's agricultural development. Nonetheless, his line has been commonly misunderstood as representing something quite different — average rainfall, rather than its reliability. Averages offer the reassurance that peaks and troughs even out over time; they convey a sense of underlying uniformity. Reliability, on the other hand, provides a measure of doubt.

The meaning of variability was perhaps even harder to comprehend at a time when science seemed destined to reveal nature's deepest mysteries. Surely Australia's curious climate was but another puzzle to be solved? As nineteenth-century scientists worked to reduce the complexity of the natural world to a series of fundamental laws, so European Australians began to suppose that their climate's capricious character might disguise some long-term pattern or regularity.[53] A thorough knowledge of these climatic cycles, it was suggested, might allow droughts to be forecast months, if not years, in advance. By the early decades of the twentieth century, the public was entranced, as various schemes for long-range forecasting were offered up by scientists and enthusiastic amateurs. Some believed that sun spots held the key, or perhaps planetary motions. Others canvassed terrestrial influences, arguing that a greater knowledge of Antarctic meteorology was crucial for the development of accurate long-range forecasts. A feature article in the *Argus* in 1938 argued that the study of both sun spot and Antarctic cycles would allow meteorologists 'to set a time-table for our weather', bringing Australia 'wealth and prosperity in boundless measure'.[54]

Most celebrated amongst the ranks of Australia's 'weather prophets' was Inigo Jones. For thirty years until his death in 1954, Jones issued long-range forecasts based on a complicated series of cycles linking planetary movements and sun spots.[55] His prognostications won a devoted following, particularly amongst pastoralists in Queensland and New South Wales. Through his own experience of drought and flood, Jones claimed he 'saw clearly the inner meaning of the problem': 'something told me that I was called to find the answer'.[56] The solar system was, he argued, 'a vast electro-magnetic machine'.[57] Once its cycles were properly understood, weather patterns both past and future would be revealed in their regularity. Within three years, Jones asserted in 1929, 'correct long range forecasts can be made giving definite areas which will be affected by drought, the period of duration of such drought, as well as giving the definite areas which will benefit by good rainfalls'.[58] Weather Bureau officials were less confident in Jones's abilities, suggesting that he submit a detailed statement of his theories for proper scientific scrutiny. H. A. Hunt admitted that some cycles showed 'encouraging' coincidences, but there comes a point, he argued, 'where the bottom falls out of any theory'.[59] The public, he warned, 'is always at the mercy of any theorist who chooses to quote statistics'.[60]

Herald, 14 January 1954, p. 4.
REDRAWN BY THE ARTIST, 2004

The possibility of uncovering the 'timetable of nature' tantalised a nation that was largely dependent on its rural industries — no more crop failures, bad seasons could be predicted and planned — but did it also appeal to our sense of rhythm? Since the Renaissance at least, the nature of time, and of existence itself, has been increasingly understood in linear and cumulative terms. The embrace of progress in western societies has dulled our awareness of the natural cycles and rhythms that underlie our onwards journeying. Yet still we find comfort in the return of spring, and stop to gaze at the moon's familiar faces.

In Chapter 3, Deborah Rose explains how Aboriginal people in the Victoria River District have developed a complex system of environmental cues foretelling the change in seasons. It is a system based not on time or a linear sequence of causation, but on connections, communication and significance.[61] There is no norm or average against which conditions can be compared. Instead, difference, absence, variability and patchiness are embraced within a rich array of local messages and meanings. The weather is never strange to those who live with it in country.

Western meteorology, however, has generally preferred causes to correlation. In the 1970s when Neville Nicholls began to examine the involvement of El Niño in Australian droughts, he found the suspicion of cycles and statistics still lingered within the meteorological community. The available evidence was 'patchy' and empirical, with limited theoretical understanding of the physical processes involved. Moreover, the very idea that climatic changes could be predicted months ahead seemed at odds with the prevailing orthodoxy. The gradual acceptance of El Niño into the meteorological mainstream has not only brought legitimacy to the field of climate prediction, it has also loosened the grip of theory and modelling over what counts as useful evidence.

Like the planetary cycles of Inigo Jones, the idea of El Niño has been embraced by the public as a sign that drought is not random or arbitrary. There is no 'timetable of nature', perhaps, but many serious droughts can be located within a recurring, natural cycle that is, to some extent, predictable. However, El Niño varies in its onset, length and severity. It is not governed by the calendar, nor read from the astronomical charts. It is a cycle that offers the possibility of prediction, even as it commits us to the fundamental variability of the Australian climate. It confronts us with a different kind of certainty, in which only constant monitoring can attune us to the global rhythms of ocean and atmosphere.

As Neville Nicholls notes, the media and the public have been inclined to overestimate the predictive power offered by El Niño. Probabilities are difficult to communicate, and the lure of certainty remains strong.[62] Despite the occasional nod to chaos theory, we expect science and technology inevitably to increase our ability to predict and control the natural world. Certainty is assumed to be the most desirable basis for rational planning. Daniel Connell describes in Chapter 7 how even a resource as obviously dependent on climatic variability as water, is fought over by primary producers demanding greater certainty in their allocations. Instead of correlations we seek causes, instead of possibilities we demand predictions. We face the challenges of variability with the expectation of certainty. And the weather continues to seem strange.

Change in the weather

The climate of the newly established colony at Port Jackson was 'changeable beyond any other I ever heard of', Watkin Tench observed, 'clouds, storms and sunshine pass in rapid succession'.[63] The weather is always changing, of course; the atmosphere is never still. Change seems natural, while unbroken spells of hot or cold weather tax our patience and understanding. At the height of summer, we scan the synoptic charts for the jagged lines that indicate a cold front is on its way — a change is coming, we sigh with relief, perhaps some rain this time. We celebrate the signs that mark the passage from one season to the next. We live our lives amidst shifting winds, varying temperatures and humidity. Change in the weather surrounds us in a way that is familiar and reassuring.

Tench was grappling with change on two fronts. While the weather continued to play its games, the colonists sought to understand and adapt. It was not the changeability of the climate that concerned him so much as the lack of reliable indicators. Without some means of prediction, without experience gained over many years, change seemed arbitrary, whimsical, perhaps even dangerous. Only with a growing knowledge of the climate could European settlers derive an understanding of the limits within which change could operate. We are comfortable with the many changes our weather brings because we have a system of prediction that moulds our expectations to the range of likely outcomes. But what if the reverse were true — if the weather itself could be changed to meet our expectations and desires? As we congratulate ourselves on the gradual acceptance of our continent's variable climate, we might ask ourselves: would we change it if we could?

Although white Australians reluctantly resigned themselves to the reality of drought, still they wondered whether the climate might somehow be changed for the better.[64] With settlement proceeding apace in the mid-nineteenth century, it was optimistically asserted that 'rain followed the plough'. A series of severe droughts prompted second thoughts, and some proposed that rainfall might actually be improved through the planting of forests. The twentieth century brought many and varied solutions, from the expansive nation-building vistas of the Bradfield Scheme, which planned to modify the climate of Central Australia by diverting water from coastal rivers inland, through to a score of would-be rainmakers targeting recalcitrant clouds with rockets, kites, balloons and guns.[65] 'Still the weather remains unconquered', the *Argus* concluded in 1944, after surveying various rainmaking attempts: 'The grim spectre of drought is one of the few enemies which man can see but cannot destroy ... with all his scientific knowledge he is powerless to kill it'.[66]

But science emerged from the Second World War with a renewed sense of potency, and, as Rod Home describes in Chapter 6, the battle against climate was taken up anew by cloud seeders in the CSIR/O Division of Radiophysics. Richard Casey, the responsible minister, proclaimed that the Australian program was 'in the forefront of research into weather modification'.[67] As the researchers prepared for a large-scale experimental program in the Snowy Mountains, the *Sun-Herald* wondered, 'Is this the year of the pay-off?' If the tests succeeded, the article concluded, 'then 1957 may go down in history, not as the year of the A-bomb, the H-bomb or the guided missile, but as the year Australia gave rainmaking to the world'.[68] A suitable gift from the world's driest inhabited continent.

Cloud seeding failed to fulfil its promise, though it has its supporters still. Neither has the vision of great nation-building projects faded from the public imagination. Variations of the Bradfield Scheme maintain popular appeal, both as an effort to thwart the menace of drought, and a symbol of national unity and achievement in the mould of the Snowy Scheme.[69] In 2002, as Australia suffered the effects of one of the worst droughts on record, media pundits led yet another chorus of 'turn the rivers inland'. Discussion focused on possible means of 'drought-proofing' the country.[70]

The experience of change raises the question of control. Do we submit to the whims of nature or resist? Do we find ways of adapting our modes of living or do we mount a technological counter-attack? In an era in which even our cars come equipped with 'climate control' and the use of air conditioning in our homes continues to spiral upwards, the answers are far from clear. In any case, the ability to effect change does not guarantee the power to control. We now know it's easy to change the weather, we've been doing it for years, perhaps since the beginning of the Industrial Revolution. After all the failed rainmaking devices and plans for drought-proofing the continent has come the reluctant acceptance that changing the climate is as simple as flicking on a light switch or driving a car — anyone can do it. Unfortunately, global warming, fuelled by the profligate generation of greenhouse gases, may well bring Australia more droughts, more heatwaves, and more 'freak' weather.

An elderly couple celebrating the breaking of the drought on Mt Fraser Homestead, Beveridge district, Victoria, 1931.
MUSEUM VICTORIA

However, as Mike Smith reminds us in Chapter 15, climate change is nothing new. For many thousands of years, human beings have adapted to shifts in the climate and conditions of the Australian landmass. Nor, as we have seen, is the idea that the climate might be changing totally unprecedented — Mr Hodgson and his fellow travellers have been issuing their warnings for decades. Why then, as Clive Hamilton asks in Chapter 16, have we been so reluctant to address the implications of global warming? Part of the answer may lie in our experience of time. Both the study of palaeoclimates and Mr Hodgson's anxieties illustrate the difficulty we have in grappling with change over time. Locating our own meagre spans within the sweep of deep time stretches our powers of imagination, just as the lessons of memory ill fit the analysis of long-term data. We remember our summers as being hotter, and yet we know that they were not. We know that the consequences of our inaction will be borne by future generations, and yet we imagine that they will still have time to spare. We guard the present against the implications of past and future. We cling to denial, Hamilton argues, when what is needed is a change in our values, and our sense of responsibility — a change in the way we think.

The papers in this volume address our experience of weather and climate from a variety of disciplines and perspectives. They invite you to find what is human in the elements and to ponder your own change in the weather.

Tim Sherratt

Tim Sherratt is a historian of Australian science and culture in the History Program of the Research School of Social Sciences, Australian National University. He was formerly Deputy Director of the Australian Science Archives Project, and has been involved in a wide range of projects relating to the history of Australian science including Bright Sparcs, an online database of Australian scientists, and the mysterious 'Cabinet of Curiosities'. More recently, he has continued his struggle to understand the meaning of the atomic age and the nature of progress in twentieth-century Australia.

2

Climatic outlooks
from revolutionary science to orthodoxy

Neville Nicholls

On Sunday morning, 7 July 1991, I was woken by two sounds. The first was the fanfare for the 6am ABC news on the motel radio-alarm clock in Dubbo — my family and I had stopped there on our drive to the Gold Coast for a vacation. The announcer led the bulletin with the news that the Bureau of Meteorology was forecasting an El Niño event, and that this would likely be accompanied by drier than normal conditions over central New South Wales. This was hardly news to me, or to my family. Two days earlier I had been involved in the regular monthly climate outlook meeting of Bureau staff at which we had decided to issue a press release about the developing El Niño and likely drought.

The other sound I heard, though, was unexpected — the sound of heavy, continuous rain. That sound accompanied us for the remainder of the drive to the Gold Coast. We arrived at the coast with the rain still falling. Very wet *Courier-Mail* billboards outside every newsagency and milk bar trumpeted:

DROUGHT FEAR
EL NIÑO RETURNS

My family thought this was hilarious — it took their minds off the question of what we could do at a beach resort in heavy, continuous rain. Luckily for the vacation, if not for farmers, the rain stopped later that day, and the remainder of the holiday (and in fact the rest of the year) was dry through northern New South Wales and Queensland.

The fact that we could issue a climate forecast, even if at times such forecasts led to embarrassment, was not something atmospheric scientists thought feasible when I started my career in 1971. This chapter describes my personal view on how scientific understanding about El Niño-Southern Oscillation (ENSO), or at least its relationship to Australia, has evolved. I will focus on the last three decades, which have seen a revolution in our beliefs about the predictability of climate.

Historical context

We need to set the scene by discussing the Australian climate and the earliest work on the phenomenon. Some of the earliest descriptions of the Australian climate noted the unusually high variability of the rainfall. William Stanley Jevons after studying only eighteen years of data noted that:

> the Australasian climate is one of irregular rains, and is thus distinguished from most climates where the rains are pretty equable and constant although less in total amount.[1]

This concept of high rainfall variability was quickly established in the Australian consciousness:

> the climate is not only niggardly on the whole, it is a most capricious tyrant, destroying at uncertain intervals what it has reared in a few milder seasons.[2]

This high variability is the result of the influence of the ENSO. Countries and areas affected by this phenomenon really are lands 'of droughts and flooding rains', to quote Dorothea Mackellar's 'My Country'.[3]

One of the areas affected by the ENSO phenomenon, with consequently amplified rainfall variability, is India. Much of the early work on climate prediction was initiated as a result of the disastrous drought-related Indian famines of the late nineteenth century. Henry Blanford, the first Imperial Meteorological Reporter to the Government of India, observed that the atmospheric pressure over the countries surrounding the Indian Ocean had been abnormally high during the Indian monsoon failure of 1877 and 1878. We now regard this period as one of the strongest El Niño events on record (I will discuss why later). Blanford requested further information on atmospheric pressure conditions at this time, from other meteorological observers around the world, including Charles Todd in South Australia. Todd was obviously intrigued by Blanford's report, and noted that Australia was also in drought in 1877. In the next major Australian drought, in 1888, Todd concluded that 'there can be little or no doubt that severe droughts occur as a rule simultaneously over the two countries'.[4] This tendency for Indian and Australian droughts to occur at the same time we now know is just one of the many 'teleconnections' that are an essential feature of the ENSO phenomenon. 'Teleconnections' are associations or linkages between climate anomalies in different parts of the world. Ångström appears to be the first to have used the term.[5] Todd's conclusion, in a newspaper article that he and two other nineteenth-century Australian meteorologists published in late 1888, appears to have been ignored or overlooked until I found it while researching for a paper on the 1888 Australian drought.[6]

Despite Todd's finding remaining undiscovered, over the next few decades several meteorologists, most notably Gilbert Walker, one of Blanford's successors in India, documented many teleconnections, named them the Southern Oscillation, and demonstrated their potential use in seasonal prediction. In Australia, E. T. Quayle showed early in the twentieth century that atmospheric pressure in Darwin could be used for seasonal prediction in south-east Australia; he simply related observed atmospheric pressures at Darwin with later rainfall over south-east Australia.

Darwin pressure is half of the Southern Oscillation Index or SOI, an index now commonly used in Australian seasonal climate prediction.[7] Quayle was, in part, inspired by Walker's work on the Southern Oscillation to look at Darwin pressures as a predictor, although he had suggested much earlier that the number of 'monsoonal' depressions over summer could be used to forecast Victoria's and New South Wales' winter rainfall.[8] Treloar, another Bureau scientist, also built on Walker's work and presented empirical evidence that total September–May rainfall in far north Australia could be predicted using Darwin pressures.[9]

There seems to have been little, if any, effort to use Quayle's (or Treloar's) work in operational climate prediction in Australia. I suspect this was because:

— The work underlying the forecast relationships was statistical, with doubts about whether these relationships would continue to hold in the future. Quayle only had a few decades of data to examine, so there was a good chance that the relationship was a coincidence.

— No one had a theory to explain the teleconnections or the lag relationships. A. M. Grant was speaking for most atmospheric scientists in the middle of the twentieth century when she warned of the 'dangers of purely empirical studies of relationships between meteorological variables. Such relationships can only be established by an increased knowledge of the underlying physical processes'.[10]

— The empirical relationships did not work very well through the 1920s and 1930s, when Quayle was testing them.

— By the late 1930s the focus of meteorologists had shifted from climate to daily weather forecasting, especially for aviation.

— From the late 1940s, as computers developed, attention shifted increasingly to weather prediction, including the development of computer models.

Throughout the atmospheric sciences in the middle decades of the twentieth century, climatology was 'neither respected nor valued'.[11] According to Kenneth Hare, 'only the old, the halt, and the infirm could be appointed to the climatological branch; the able-bodied men were expected to be forecasters'.[12] Climatology was regarded as 'mere book-keeping ... to be posted to the climatological branch of a national weather service was like being made an intelligence officer, or a lighthouse keeper; it was a terminal appointment'.[13]

As a result of the shift in focus away from climate science, in the nine-month training course for meteorologists that I completed in 1971, the Southern Oscillation was dealt with in less than ten minutes. And my lecturer said that all it meant was that when pressure was high here it was low somewhere else and that 'it wasn't very important'.

One way to highlight the shift in emphasis in meteorology is to contrast two public comments, made about sixty years apart, on the feasibility of climate prediction. The first was published by Ellsworth Huntington, the geographer and environmental determinist, who visited Australia in the early 1920s and asked:

> Why should not Australia find out how to predict the general types of weather six months or a year in advance? ... That this can be done and that the general character of the weather for six months or a year in advance can some day be predicted with considerable accuracy seems certain. It is merely a question of

securing able men, giving them every possible facility for investigation at home and abroad, and permitting them to study their problems for years without being forced to show immediate practical results.[14]

Sixty years later we no longer had Huntington's confidence, perhaps because we had not been provided with 'every possible facility for investigation'. At a 1983 meeting on El Niño's impact on Australia, a senior Bureau scientist was asked why the Bureau did not issue forecasts based on the statistical relationships found by Quayle and others. In reply he pointed out that we needed data from all over the world to make forecasts for just five days ahead, so the idea that a single indicator (the Southern Oscillation Index) could be used to forecast a season ahead was just not credible. This was the orthodox view about climate prediction amongst meteorologists in the 1970s and early 1980s.

Attempts had been made to prepare forecasts out to a month or more, both in Australia and overseas. Kenneth Arrow, Economics Nobel Laureate of 1972, was a weather officer in the US Air Force during the Second World War. Some officers had been assigned the task of forecasting the weather a month ahead, but Arrow found that these forecasts showed no skill. The forecasters agreed with his assessment and asked their superiors to be relieved of this task. The reply was 'The Commanding General is well aware that the forecasts are no good. However, he needs them for planning purposes'.[15]

In Australia, the Bureau of Meteorology experimented with thirty-day forecasts in the 1960s. These forecasts extrapolated the observed broad-scale circulation fields around the hemisphere (that is, they did not use any information about the ENSO), and used these extrapolations to estimate temperature and precipitation for the ensuing month. The forecasts showed little if any skill and were discontinued at the start of the 1970s.[16]

Charles Todd, the South Australian Superintendent of Telegraphs, was one of the pioneers of what we now call the El Niño-Southern Oscillation. In 1888 he recognised that Australian and Indian droughts tended to coincide. This 'teleconnection' is one result of the Southern Oscillation.

BUREAU OF METEOROLOGY

Despite the widespread belief that climate was fundamentally unpredictable, some researchers continued to study the now unfashionable Southern Oscillation.[17] By the 1960s, Hendrik Berlage, working in what was then the Dutch East Indies and in the Netherlands, had recognised that the Southern Oscillation was related to the temperature of the east equatorial Pacific Ocean. Jacob Bjerknes demonstrated how El Niño events could produce climate teleconnections. Since then, the Southern Oscillation and the El Niño have been lumped together as ENSO. Within Australia, Sandy Troup at the CSIRO Division of Meteorological Physics wrote one of the classic papers on the Southern Oscillation.[18]

The modern era

By the time I started working in the Synoptic Research Branch of the Bureau of Meteorology, a substantial amount of documentation of ENSO was available but tended to be hidden in obscure reports or occasional journal articles. My first task in the Bureau was that of the most junior member of a team studying a severe drought in the soon to be independent Papua New Guinea. This drought was most severe in the highlands between June and September of 1972. Our report was prepared speedily but did not suggest a cause of the drought. Not until the following year (much too late to be of any value in prediction) did I realise that the El Niño of 1972 had been a major influence on the Papua New Guinea drought.[19] The main reason for this delay in attributing the 1972 drought to El Niño was the slowness of data collection, especially from the oceans — we simply did not know that an El Niño was under way, while we were studying the drought that it was causing. But neither had we been looking for an El Niño — a meeting on the development of long-range forecasting in the Bureau of Meteorology, chaired by the then Director of the Bureau, W. J. Gibbs, in February 1972, did not even mention ENSO.[20] El Niño was not on anybody's radar. The historical impact of ENSO on Papua New Guinea has since been comprehensively documented.[21]

Soon after the 1972 El Niño and drought, the idea that the Southern Oscillation was an important control on Australia's climate was documented by Barrie Pittock and Peter Wright.[22] There was still widespread scepticism amongst atmospheric scientists about the feasibility of seasonal climate prediction. Bruce Morton, at the end of a conference on climate change and variability held at Monash University in 1975, noted that 'to predict when seasonal and annual forecasts will form a useful part of our meteorological services is itself a long-range forecast, and at present beyond our skills'.[23] However, Wright, at the same conference, did note that calculation of the lag relationships associated with the Southern Oscillation would 'greatly improve prospects for seasonal forecasting'.[24]

After my experience with the New Guinea drought and El Niño I was convinced that we could use the Southern Oscillation for seasonal prediction. When I stumbled across the early work of Quayle, I realised we now had sixty years of independent data, not available to Quayle, to confirm that Darwin pressures (or the Southern Oscillation Index) could be used to forecast east Australian rainfall from about the middle of the year.[25] The new, independent data also allowed confirmation that Indonesian early wet season rainfall could be predicted, as Cornelis Braak had suggested over sixty years earlier.[26] I also showed that the date of onset of the Australian wet season and seasonal tropical cyclone activity around Australia could be predicted using the Southern Oscillation Index.[27] The tropical cyclone paper was reviewed by Bill Gray from Colorado State University who went on to demonstrate that Atlantic hurricane activity could also be predicted using ENSO. Subsequent work by many researchers has demonstrated that the phenomenon affects tropical cyclone activity in all the ocean basins, and forecasts are now available on the internet.[28]

Tentative attempts to forecast an early onset to the North Australian wet season were made in late 1981, based on the relationship between onset date and ENSO. A forecast was released to the media in Darwin, but the Bureau of Meteorology decided not to implement an SOI-based forecast system at that time, for reasons quite similar to those I suspect stopped the use of Quayle's work earlier. The two main reasons were because the system was 'patchy' (that is, it only worked for some seasons and regions) and because the relationship and why it could lead to climate predictions were not understood. The work was purely empirical, with no theoretical understanding of the causes of these relationships. Meteorologists, whose work now relied on the dynamic prediction of weather with a strong theoretical underpinning, were suspicious of a complete reliance on empirical results.

By the time the major El Niño of 1982 appeared, there had been some improvements in data collection, and it had been realised that the El Niño was the product of tropical Pacific air–sea interaction. Some scientists had even found what appeared to be useful clues to the onset of an El Niño. Satellites now provided global sea surface temperatures. However, the eruption of El Chichon contaminated these measurements. As late as October 1982, some Northern Hemisphere climatologists were convinced that there would not be an El Niño in 1982–1983. In Australia we were concerned about the very high Darwin atmospheric pressures apparent from May 1982. Such high pressures had been a clear sign of an El Niño event in the past, so it seemed to us by mid-year that a major El Niño was under way with predictable consequences.[29] At a September 1982 seminar with the politically incorrect title of 'Farmers and the Weatherman', I observed that we were unlikely to get drought-breaking rains until well into the next year since El Niño related droughts usually lasted into the New Year. I used the onset of the El Niño that year as an excuse not to plant a garden around my new house. There were more important impacts of the El Niño, however. Close relationships were established at this time between Queensland agricultural scientists Greg McKeon and Graeme Hammer and climate researchers, partly because of the public statements that the drought was likely to persist into 1983. These relationships have led to Australia being the world leader in climate forecast applications, because we have been able to combine meteorological knowledge with an understanding of agriculture.[30]

Around this time a colleague, John McBride, suggested that we should write a paper describing the seasonal relationships between ENSO and Australian rainfall. I was not enthusiastic — I thought this had been adequately described and documented, and was generally understood. But John pressed me and so I insisted that he should be the lead author, partly because he had suggested the idea and partly because I did not think the paper would be cited very frequently. I was wrong — the paper probably has been the most cited paper for both of us.[31]

I was also wrong about how widely accepted the relationship was between ENSO and Australian rainfall. Many, perhaps most, meteorologists were still suspicious. In March 1983 the Australian drought broke, while the El Niño was still strong. A senior Bureau meteorologist told me that this demonstrated that there was no real relationship between El Niño and Australian rainfall. In fact Australian droughts often break while the warm sea surface temperatures of the El Niño are still in existence in the east equatorial Pacific — in many cases in the past the breaking of the Australian drought has been the first sign that the El Niño was about to collapse.

The scepticism of many atmospheric scientists about the use of El Niño in climate prediction was still strong, partly because there was still no fully developed theory of the phenomenon, and no computer model had simulated the phenomenon. The potential use of ENSO in climate prediction still rested on empirical relationships found by Quayle, Walker and others, although these had now been repeatedly confirmed by decades of new data. Another reason for the reluctance to issue climate forecasts was the fact that the relationship between ENSO and Australian rainfall had varied somewhat over time.[32] In the absence of a detailed physical understanding of the phenomenon, the only way to see if this was likely to be a major problem was to find more data. Useful rainfall records for Australia did not go back much further than the late nineteenth century, although we had good dates of El Niño events back for 500 years (from documentary evidence of flooding on the Pacific coast of South America). One day when I grumbled about this lack of early data to my wife Jill (a librarian for ten years in the Australiana section of the State Library of Victoria), she told me to check the reports from the eighteenth and nineteenth-century governors of New South Wales to the colonial secretary in London. I searched these reports for mentions of drought and found that most of the droughts from 1788 to 1841, that is, up to the instrumental era, were associated with El Niño events.[33] So, even if there was some variation in the relationship, it had been generally consistent for over 200 years.

I needed more data than this, however, to argue that the phenomenon had been affecting this country for a very long time and would therefore, presumably, continue to affect us. I started to argue that the well-known adaptations of the Australian biota to highly variable rainfall suggested that ENSO had been operating and amplifying climate variability in Australia for thousands of years.[34] This argument has now received some support. Tim Flannery, for instance, suggests that 'Judging by the way the Australian flora and fauna have adapted to it [ENSO], I suspect that it has been in operation for very much longer [than the historical climate record]'.[35]

When I presented this argument at a symposium at the University of Melbourne in 1987 I added the claim that highly variable climates would make permanent agriculture less attractive and thus hunter-gatherer societies might have survived longer in areas affected by ENSO. I pointed out that those hunter-gatherer societies that did survive into the nineteenth century were indeed located in areas affected by ENSO. Anthropologists at the symposium labelled me a 'climate determinist' for raising this suggestion. In the face of this determined attack, I never submitted the hunter-gatherer argument for journal publication, although it remains in the symposium proceedings.[36] Nevertheless, I was convinced that the adaptations of the Australian fauna and flora to the highly variable rainfall meant that ENSO had been operating for a very long time, and that it should continue to be useful in climate prediction. The argument that the irregular rainfall associated with ENSO would make agriculture a less attractive way of life seems a more acceptable argument nowadays, at least for Australia.[37] Tim Flannery notes, for instance, that 'It makes an enormous amount of sense to me to see the lack of agriculture by Australian Aborigines as a fine-tuned adaptation to a unique set of environmental problems'.[38]

My interest in the various adaptations to the highly variable climate associated with ENSO arose because I had been trying to develop methods for the direct prediction of impacts of the above phenomenon on crops, human health (Australian Encephalitis), and native fauna and flora. In the mid-1980s Col Limpus from the Queensland National Parks and Wildlife Service contacted me regarding observations of large year to year fluctuations in the numbers of breeding green turtles. Some of his colleagues thought this might reflect variations in Indigenous hunting of the turtles, but Col thought they might be related to ENSO and it proved to be so. This meant that we could predict, well in advance, the numbers of breeding turtles (useful for management purposes) and also dismiss the theory that the variations were the result of direct human action.[39] This was one of several examples I used to demonstrate that ENSO could be used to make direct predictions of impact variables such as crop yields without worrying too much about predicting the intermediate climate variables.[40] I saw this as a means to side-step the reluctance of meteorologists to initiate seasonal climate prediction using ENSO.

Operational climate prediction

In 1986 the success of the first computer model forecast of an El Niño changed the situation, and meant that my efforts on impact prediction and the historical and pre-historical relationship between drought and El Niño suddenly became more than interesting sidelights. The forecast, by Mark Cane and Steve Zebiak, was received suspiciously by many scientists, including myself, but it proved to be correct. This successful forecast made ENSO forecasting respectable in the eyes of most meteorologists, and provided a further boost to research on the phenomenon. By this time, we had worked out some of the details of how the ocean and atmosphere interacted to link the atmospheric Southern Oscillation and the oceanic El Niño. This helped legitimise the phenomenon in the eyes of physical scientists, leading to enhanced interest in its use in prediction.

Media and political interest in the phenomenon also increased. After an article about El Niño and its impact on Australian crops appeared in the Adelaide *Advertiser* in early 1988, Senator Robert Ray, representing the Minister for Administrative Services, answered a question without notice about seasonal forecasting. The question alluded to an article in the *Advertiser* claiming that Bureau of Meteorology funding was insufficient to allow El Niño monitoring and prediction. Ray stated that 'The Bureau's ability to monitor these climatic fluctuations and its scientific understanding of phenomena such as El Niño and the so-called southern oscillation have reached the stage where it expects soon to be able to begin issuing regular public advice through the media on the general state of the climate system over Australia and the likely seasonal outlooks'.[41] Mike Coughlan, from the Bureau's National Climate Centre, and I had been preparing and revising a proposal for the forecasts for over a year at that stage. Through 1988 we prepared trial seasonal outlooks, but these were not released to the media (they were stamped 'Confidential'). Then a *New Scientist* article on 25 March 1989 criticised the Bureau for not releasing warnings of the likelihood of widespread wet conditions. Severe

A drying waterhole in eastern Australia during the 1982–1983 drought.

BUREAU OF METEOROLOGY

flooding had followed these heavy rains. The article observed that the Bureau had noticed the similarity of the current climate situation to previous widespread flooding events 'but feared that they might be wrong and so made no public warnings'. Soon afterwards the seasonal outlooks started to be distributed to the media. These forecasts were still based on the statistical relationships dating back to Quayle's early work.[42]

The El Niño event of 1877–1878

Now I want to return to the 1877–1878 El Niño event — the event I noted earlier as one of the strongest on record. We only know this because of a series of fortuitous connections. The Southern Oscillation Index is one of the main indicators for comparing strengths of historical El Niño events. The Southern Oscillation Index uses atmospheric pressure data from Tahiti and Darwin. The Tahiti data were available from 1877, but early in the twentieth century the Bureau of Meteorology seems to have destroyed the Darwin observations from before 1882, because of doubts about their reliability. I stumbled across the Darwin pressure data for 1876–1878 in a volume of meteorological observations made in South Australia and published by Charles Todd.[43] I mentioned this to Jill, now a Bureau librarian but still with a professional interest in Australian historical documents. That same day another Bureau librarian had shown Jill a list of old documents dealing with meteorology that he had found in a rarely opened drawer in the Library. Jill realised that one of the documents listed what appeared to be the original observations journal for Darwin, and that the list was annotated with what looked like archivist's codes. Jill suggested we look for the journal in the Australian Archives, and indeed that was where we found it, in Darwin. Ian Butterworth from the Bureau's Northern Territory Regional Office painstakingly copied the daily observations (the journal was too delicate for photocopying) and Rob Allan from CSIRO checked and corrected them. We now had the original observations and could calculate the Southern Oscillation Index for the 1877–1878 El Niño event and compare the 1877–1878 and the 1982–1983 events.[44]

The El Niño event of 1982 and its consequences

The very strong 1982 El Niño led to a major improvement in observations. The equatorial Pacific Ocean was now instrumented with buoys that reported sea surface and sub-surface temperatures via satellite almost instantaneously. By the 1991 El Niño, we were monitoring the east Pacific in 'real time', allowing the forecast I mentioned at the start of this chapter. In 1994, when the next El Niño developed, we issued a press release predicting an increased likelihood of drier than normal conditions. This time the forecast was issued just as a major rain-bearing system drifted over south-east Australia. Even as we made the forecast we knew we would be lambasted by colleagues, friends and the media for talking about a drought while the rain pelted down. But the rain lasted less than a month, and the remainder of the year was dry, as expected.

Through the 1990s, the public, media, and meteorologists became more familiar with ENSO and its potential for prediction. Much of this increased familiarity was the result of efforts in Queensland initiated by Graeme Hammer and Greg McKeon. With Roger Stone, an ex-Bureau of Meteorology employee, Jeff Clewett, and many others, they worked to raise the level of understanding and application of the phenomenon throughout Queensland agriculture. Nationally, Mary Voice, from the Bureau's National Climate Centre, David White, Bureau of Resource Sciences, and Barry White, coordinator of the Climate Variability in Agriculture Programme, all worked to ensure that the public could access forecasts, and could understand them.[45] These forecasts are quite complicated, so presentation was important, if we were to achieve the best possible use.

The importance of the presentation of the forecasts became very clear during 1997. Our confidence in our predictions had increased with the 1997–1998 El Niño. The forecasts made in 1991 and 1994 had contributed to this. By the 1997 event, we were also more confident in our understanding of how the forecasts should be presented. Because the climate has a strong 'chaotic' element to it, even in areas where ENSO is a major controlling factor, forecasts should not be expressed in deterministic terms, such as 'we predict a drought', but instead, 'the existence of an El Niño shifts the probability distribution, making drier than normal conditions more likely than in other years'. Our forecasts, by the 1997 event, were expressed as probabilities, and we thought we understood how people would interpret these forecasts. When parts of eastern Australia received good rains in spring 1997, we were surprised at the vehemence of the attacks on us and our forecasts in the media. Our forecast for spring 1997 indicated that there was up to a 66 per cent chance of rainfall being in the lowest historical tercile (compared with a climatological probability of 33 per cent). So, although the prospects for 'dry' conditions were doubled by the existence of the El Niño, this did not seem to justify the attack as a 'wrong' forecast when spring rains turned out to be average-to-good. However, the flood of news of El Niño related disasters from around the world appeared to push forecast users to expect the worst. So we read articles criticising the forecasts, with headlines such as 'El Niño fails as a reliable indicator'.[46] We learnt from this experience that even with great care it is difficult to express probabilistic forecasts in a way that would be interpreted and applied as we expected. So I searched the cognitive science literature for ways to improve the presentation and dissemination of the forecasts.[47] We are still some way from getting the message across correctly. A journalist on the Channel 10 News on 18 July 2002, discussing the drought situation around Bourke, stated that there was 'no rain predicted here until March next year'.

During the 1997–1998 El Niño there were El Niño cartoons, El Niño jokes, El Niño posters. Articles about it appeared in *Time* magazine, and many other general publications. Taxi drivers, journalists, my children's teachers (to my children's embarrassment) — everyone wanted to talk about the El Niño. It is interesting that even though there had been many attacks on the El Niño based forecasts during 1997, by the time forecasts of a 2002 El Niño started appearing, even more organisations and media outlets were publicising, and sometimes exaggerating, the potential consequences for Australia. During winter 2002 it was difficult to read a newspaper, even the business pages, or watch a television news bulletin without seeing a reference to El Niño. This partly reflects how in Australia we seem now to have equated El Niño with drought, even though not all droughts are related to El Niño events, and vice versa.[48] So the eastern Australian drought of 2002 was destined to be linked, in the media and public psyche, to El Niño. This has been a remarkable change in three decades — from a time when meteorologists could ignore the phenomenon and its potential use in prediction. Other countries, including the United States, now also use ENSO in seasonal climate prediction. The successful use of the phenomenon in climate prediction in Australia encouraged its use elsewhere.[49]

Another change is that other climate signals, such as the Antarctic Circumpolar Wave (ACW), are now trumpeted as ways to improve our predictions. Two decades ago it seemed impossible to convince our colleagues of the reality of ENSO and its use in prediction. Now it is all too easy to suggest a new 'predictor' that will, so it is claimed, provide more accurate predictions than ENSO. One reason why this has become easier is the great improvement in the availability of global historical climate data sets, such as precipitation, and the availability of computers to analyse these data. Three decades ago the collection of appropriate data and a search for ways to predict climate variations would require several people, large computers and some months. Nowadays, these calculations can be done on the internet in a few minutes using other people's software, computers and data sets from anywhere in the world. What once might have taken three months and could only be done with extensive computing and technical support, can now be done in three minutes at any internet café, anywhere in the world, by anyone.[50]

The availability of such resources, and their 'democratisation', is one reason why 'new' climate phenomena are being postulated with increasing frequency. The speed with which new climate phenomena are now proposed to enhance our climate predictive skill contrasts with the slow acceptance of ENSO as a predictive tool. As I have noted earlier, one of the reasons for the slowness in the acceptance of ENSO was that for many decades there was no theory explaining its characteristics. Nor was there, until recently, a computer model capable of simulating, let alone predicting, its development. The empirical evidence supporting its use in prediction was very strong and well documented, but a meteorological community committed to theory and modelling as a basis for belief did not consider this sufficient. Once theoretical concepts and modelling studies were available to underpin the empirical evidence of ENSO, its acceptance was quite rapid. In turn, the success of the phenomenon in prediction has, I believe, opened the door to the acceptance of other climate 'predictors' based purely on empirical evidence. This may not, however, be without its dangers — purely empirical results can be flukes. This is almost certainly the case for some of the recently proposed climate predictors.

Conclusions

The story of the development of the science of El Niño to the point where it is used routinely in operational forecasting provides a classic example of how scientific 'facts' develop. Trenn and Merton point out, paraphrasing Ludwig Fleck, that 'Discovery in science, whether modification or transformation, and whether of a theory or its thought style, is a complex, socially conditioned product of collective effort'.[51] A complex of interactions and collective effort, including theoretical underpinning and computer simulations, as well as much empirical work, was needed to allow the atmospheric science 'thought collective' to accept the feasibility of climate prediction. The lack of these elements, and the belief that the atmosphere was fundamentally chaotic and unpredictable at long lead times, formed what Fleck called a 'structurally complete and closed system of opinions' which, earlier, offered resistance to the notion of climate outlooks. No single individual, or even a group of individuals, was responsible for reversing the belief that climate was unpredictable. Rather it was the work of many scientists over a considerable period that led to the change in the 'thought collective' to the point where, now, many in the public and media believe the climate is more predictable than is actually the case.

Neville Nicholls

Neville Nicholls is a Senior Principal Research Scientist in the Bureau of Meteorology Research Centre and leads the Climate Forecasting Group. He has been studying the impacts of El Niño – Southern Oscillation (ENSO), and its use in prediction, for over thirty years. He has published widely on a variety of topics including ENSO, statistical techniques in climate research, and climate and history.

Untitled bark painting by Nakkurridjilmi Nganjmira showing Namarrkon, the Lightning Spirit, and dancing mimi spirits warning of the impending wet season with its accompanying storms

NATIONAL MUSEUM OF AUSTRALIA
PHOTOGRAPH BY MATT KELSO

Change of Seasons

This bark painting depicts Namarrkon, the Lightning Spirit, and dancing mimi spirits. The Lightning Spirit is shown with conduits of power running from his large testes to his hands. Stone axes, representing the lightning strikes that smash trees, protrude from his knees. To his right, a ceremonial dancing ground is delineated by a black band and red crosshatching on white. One mimi plays a didjiridu, another plays clapsticks, while another sings. The others, including one holding a spearthrower, are dancing.

Informants say the painting is a warning to people to finish ceremonies before the time of year when lightning is common. Apart from funerals, the great majority of ceremonial activity in western Arnhem Land occurs in the dry season. It is said that men can use power songs to direct lightning activity to the camps of enemies.

3

Rhythms, patterns, connectivities

Indigenous concepts of seasons and change

Deborah Rose

The Canadian anthropologist Julie Cruikshank has made an important study of the Canadian Yukon people's knowledge of glaciers, attempting to link that knowledge with current scientific research, particularly in relation to global climate change. She notes that much of the literature on traditional ecological knowledge (TEK) tends toward classifying knowledge as data, and toward incorporating data into existing western paradigms. More attention should be given, she argues, to engaging with local knowledge as expertise that could contribute to theoretical perspectives.[1] In this chapter I take up that challenge, but not, of course, with respect to glaciers.

Cruikshank provides a marvellously economical statement of Yukon world view in relation to landscapes: human–nature relationships, she states, are about sociality and sentience.[2] In an Australian Aboriginal context, these concepts of sociality and sentience are equally insightful, distilling major propositions concerning the nature of human and non-human life. To these two concepts we must add connectivity. In both societies — Canada and Australia — the question can be posed: can a world view that posits climate or weather (or glaciers) as sentient, social and connective find common ground with western science?

In one very important sense it does not matter whether there is common ground or not. The search for common ground can very readily degenerate into an imperialism such that other people's ideas are accorded value to the extent that they mesh with our own. Rather than appropriation, however, I aim for dialogue. I believe that we need to have cross-cultural conversations in this period of global climate change and probable imminent crisis. In this chapter I explain why.

Sentience, sociality, connectivity

Cyclone Tracy came off the sea and slammed into Darwin on 24 December 1974, one of the most holy Christian nights of the year. It crossed Darwin, turned around, and slammed it again. Its energy, its focus on the major site of white Australian habitation in the Far North, and its timing, all mark it as something out of the ordinary. As Tumarkin notes, it was the first cyclone to be interpreted as a *national* disaster.[3] Holland and McBride provide a very comprehensible account of tropical cyclones, including Tracy.[4] Their analysis is devoid of human factors with the exception of noting human vulnerability with increasing populations in cyclone prone areas.

Why did Cyclone Tracy hit Darwin? Local Larrakia people had an answer: in their view, Cyclone Tracy was a hit against whitefellas who were refusing to grant the Larrakia people land rights.[5] In other words, the cyclone had a social context, and it came to a particular place — Darwin — because that was where the problems were. Across the region affected by Cyclone Tracy, the cyclone has become part of Aboriginal people's repertoire of stories. Other stories told elsewhere also identify a social context for this extreme event, but not all focus on the Larrakia. Over in the Kimberley, for example, they made a whole corroboree about it. Krill Krill tells the story of the Cyclone through song, dance, body paint, and headdresses. It is one of the most vivid productions in a region where Indigenous culture is rich with corroborees. Krill Krill was taught to Aboriginal artist Rover Thomas by his mother a few months after she died. She told him of her travels, how she had seen what the Cyclone had done to Darwin, and how she wanted the story remembered by being performed. Aboriginal elders in the Kimberley attribute this meaning to the cyclone: that Aboriginal people, young and old, should take it as a warning to keep their knowledge and culture strong.[6] Thomas's work hangs in the National Gallery, and some of the paintings depict Cyclone Tracy.

Earlier generations of secular Westerners would probably have dismissed these stories as fanciful or primitive attempts to explain that which Indigenous culture could not explain in more scientific terms. Such stories were thought to arise from a lack of rational analysis. Anthropologists might have dubbed some of these explanations 'proto-science': endeavours to assert a sense of control in the absence of powerful technology.

Today, I think we are all more cautious. Not only is technology a power that generates problems as well as controls them, but that positivist sense of the separation of humans from nature is wavering. Surely if a butterfly in an African jungle is causally connected to a snowstorm in New York, it is not out of the question that human action may be causally connected to a cyclone. Chaos theory would tell us this, by implication at any rate. I am proposing a very modest reading of context that includes humans as participants in the flux of the world.

Seasons in the Victoria River District

My discussion of seasons is set in the Victoria River country of the Northern Territory, in the Aboriginal community of Yarralin, excised from Victoria River Downs station, and draws on earlier published work, especially my ethnography, *Dingo Makes Us Human*.[7]

Here in the savanna region of the monsoonal north, white settlers established broadacre cattle stations in the Victoria River District just over one hundred years ago. Overrunning the homes of the Indigenous peoples of the region, they first shot and hunted the local peoples, and later pressed them into service on the cattle stations as an unfree and unpaid labour force. In the mid-1960s Aboriginal people in this region went on strike against the appalling system of oppression that ruled their lives, and as they gathered support throughout the nation, they began to articulate the underlying purpose in all their forms of resistance: to regain control over at least some portions of their traditional homelands. From the 1967 referendum, which gave Aboriginal people a stronger moral claim on the Commonwealth government, people participated increasingly in national and international struggles for equity and cultural survival.

In the inland arid to semi-arid tropics, conversations about weather tend to focus on rain: will it ever come, and later, will it ever stop? According to my Aboriginal teachers, sun and rain are the two great meteorological forces. A consideration of seasons thus starts with sun and rain. Since any temporal starting point is arbitrary, I begin in mid-year of the Anglo-Australian calendar, which is also mid-winter in the temperate zones of the southern hemisphere, and mid-dry in tropical Australia. This is the cold time of year; the sun is working to slowly heat up the earth. The earth becomes hotter and hotter, with increasing temperatures and humidity. Toward the end of the year, or early into the following year, the rain breaks up the intense heat. With the rains, the rivers start to flow again, the ephemeral waters are replenished, green grass grows, and life starts to flourish. By the end of a good wet season, in March or so, the country is well-watered and lush. The end of the rain is brought about by the big winds that bring cold weather. The sun gains ascendancy and begins to dry out the earth and the waters, and to warm the earth that has been cooled by rain and wind.

In a later section I will discuss some seasonal indicators, but for the moment I stay with the broad story. Why does the rain come? My teachers said that it comes because the flying foxes tell the Rainbow Snake to get up and work. In the cold time of year the flying foxes are in the 'top country' away from the rivers. As the sun dries the country, they move toward the river, and hanging in the trees over the river, they call to the Rainbow Snake to rise up and bring rain. If you think these are fairy tales, I would like to ask you to put that thought aside and focus on connection and communication. The stories I present from my Aboriginal teachers are in one sense neither true nor false. They are human ways of communicating about a communicative system formed through connectivity. That is what I want to communicate, without destroying the liveliness of the stories.

The story can be fleshed out, because while there are Dreaming origins for the relationship between Rainbow and flying foxes, and while these relationships articulate some of the main social categories that bind human and animal species into groups of shared flesh, there is an ecological side to the story as well. Flying foxes feed, by preference, on the blossoms and nectar of eucalypts. Yarralin people point especially to the inland bloodwood (*Eucalyptus terminalia*) and the magnificent tree known in vernacular English as the half bark (*E. confertiflora*). Both of these species produce large, showy and heavily scented flowers, so they are obvious candidates for both flying fox and human attention.

In the Victoria River region, eucalypts flower in succession from higher ground to lower ground, which is also to say from the drier country on the hillsides down to the river banks and channels. My teacher, Jessie Wirrpa, divided the eucalypts into those which flower in the dry time and those which flower in the rain time. *E. terminalia* (*jartpuru*, bloodwood) and *E. confertiflora* (*ngurlkuku*, half bark) are among the prominent dry time flowerers, along with several other species including *E. pruinosa* (*yarirra*, smoke tree) which grows out on the dry flats and up the lower reaches of stony hills, and *E. microtheca* (*wulwaji*, coolabah) which grows around billabongs. *E. camaldulensis* (*timalan*, river red gum) is the outstanding example of those which flower in the rain time, and the river melaleucas flower then as well (*pakali*, *Melaleuca argenta* and *M. leucadendra*). Along a big river like the Wickham, the banks are lined with paperbarks and river red gums. They burst into flower just before the rains come.

The flying foxes follow their preferred food, and it brings them to the riverside just before the rains where they forage in the thousands. Yarralin people say that the flying foxes talk to their mate the Rainbow, telling it to move, to get up, to get to work, to bring the rain. They say that the earth is getting too hot, that everything is too dry. We see here a sequence of communicative events: the flowering of the eucalypts is triggered by each species' own internal 'time' and its response to local conditions. The flowers are food for the flying foxes, who then follow their blooming. The presence of the flying foxes at the river to feed on *E. camaldulensis* thus constitutes an event that signals a particular moment of seasonal succession: there will not be any more eucalyptus flowers until the rains come and revitalise the country.

Flying foxes are joined by a chorus of others, all telling the Rainbow to move and act. The Rainbow listens and rises up; it towers over the earth, emitting lightning, thunder and rain. People, too, may add their songs to the multitude of voices demanding rain. The first rains start to create the moisture that moves back into the sky, forming more clouds and rain.

The rain cools the earth, and the sun loses its ascendancy. Now there is the time of year called *mayiyul* in Ngaliwurru language when it rains and rains. *Mayiyul* builds on a root denoting the world of plants. The end of the rainy time, like the beginning, is brought on by an accumulation of messages. A little bird (species unidentified) sings, telling it to go away. At the same time, the cold wind comes. It rises up to wrestle with the Rainbow and to send it back down into the water. People may sing their songs for the wind to send the Rainbow back into the depths of the permanent water. Dry and cold weather return.

One way of thinking about seasons is through the contrast between wet and dry. This contrast pulls the earth into an interactive dynamic: water and clouds form the wet side of the story, dry ground and sun form the dry side of the story. Rainbow and sun, wet and dry, oscillate back and forth through a dynamic that is both agonistic and synergistic.

The contrast between sun and rain, while clear and unambiguous in many contexts, is by no means absolute. Rather it is cross-cut by other contrasts. Another way of thinking about seasons develops around the hot/cold contrast. Hot and cold times are indicated by constellations. The Pleiades, or seven sisters (*jarinkarin*), tells cold weather, and a constellation sometimes called 'seven' because of its shape (unidentified as yet), tells the hot time. When this cluster of stars gets high in the sky, it means that hot weather has fully arrived. Although the constellations refer to broad shifts in the earth's orbit, and not necessarily to local phenomena, they provide a clear template for identifying this contrast.

The hot/cold contrast neatly intersects the wet/dry one: the sun works as the source of heat during the cold period whereas the rain cools the earth during the hot, wet season. One effect of the contrast between hot and cold is that the apparent absolute difference between sun and rain is cross-cut by another set of contrasts which shows the one building up out of the other in a more complex fashion than is suggested by a contrasting binary. Sun and rain are thought of as opposing forces, but on the other hand, and not contradictorily, rain can and does fall at any time of year, and the sun shines, even if only briefly, on almost every day of the year. The two forces interpenetrate each other; their struggle for ascendancy derives as much from their entangled mutuality as from their oppositional autonomy.

Calendars

The monumental work edited by Eric Webb and published in 1997 — *Windows on Meteorology* — contains a number of superb essays on ethno-meteorology. I draw primarily on the chapters in this book in a discussion of calendars.

There are many efforts in existence today to situate Aboriginal seasons within a western calendar with its twelve months and four seasons. As Jane Simpson notes, there is a large demand for such calendars for use in schools where bi-cultural education is desired and promoted.[8] She notes several problems with them, and each is worth flagging. One is that non-Aboriginal teachers run the risk of imposing Eurocentric ideas onto seasons that are not appropriate (the term 'winter' for dry season, for example). Another is that calendars may be thought to apply more widely than they do, leading to children being taught 'indigenous' knowledge that is at odds with and possibly undermines the Indigenous knowledge of their own people and their own country. A third problem is addressed more directly by Robert Hoogenraad and George Jampijinpa Robertson in their article on central Australia. They ask whether the use of calendars is even appropriate. That is the point I will pursue, looking first at questions of 'fit', second at questions of imperialism, and third at those of loss.

Hoogenraad and Robertson propose that a major difficulty in getting central Australian Indigenous seasons to fit a Western calendar is that 'arid Australia does not have truly cyclical seasons'. They discuss a region in which Indigenous meteorological knowledge focuses on the connections between weather pattern indicators and the availability of water, flora and fauna. They note the significance of features that include temperature, winds, clouds, length of day, constellations and other factors.[9] Much of their analysis accords closely, at a general level, with my account of Victoria River seasons. Hoogenraad and Robertson go on to suggest that the central Australian calendars are very different to those of the Top End monsoonal regions, and they query the purpose of 'squeezing the very variable Central Desert seasons into a chart at all'.[10]

There is a certain compelling logic to the idea that there may be greater and lesser degrees of variability and that where there is greater predictability, Indigenous knowledge may be more readily mapped onto a Western calendar. I think that is probably accurate, but it does not assist in the question of whether we should be doing this at all, and it may promote a stronger statement of Far North predictability than is actually the case.

Imperialism leads us to another issue: a Western calendar is a Western artefact. What happens when another meteorological world view is squeezed into it? The calendars produced in the Top End do appear to represent a relatively predictable system. But even across the Top End there is no agreement on how to represent ethno-meteorology. Stephen Davis's work is widely used in this context.[11] He takes an extraordinarily concrete approach, mapping seasons almost as if they were geological strata. In contrast, Betty Meehan and the late Rhys Jones, working in the same part of Arnhem Land as Davis, produced an account that differs from his in almost every way. They start with Indigenous seasons defined by winds, of which there are two major ones — the north-west wind that brings the monsoons, and the south-east winds that bring the dry. These two winds were said by Meehan's and Jones's Aboriginal teachers to be 'in opposition to each other, alternatively occupying the centre stage. Their relationship is likened to that of trading partners exchanging important ritual gifts and songs.'[12]

Jones and Meehan do not attempt to map the winds and other broad features onto a calendar. The system they present is one of action: the big forces of weather do their work in the world, and people along with other living things work with those patterns. In an earlier article, however, Jones discussed his teachers' classifications of specific relationships, temporally sequenced within the broad patterns.[13] The Davis calendar is, by contrast, static. Everything is mapped into its place. Motion is depicted as clockwise extension around the circle; but who, or what, is moving?

Equally importantly, Jones and Meehan note that while some of the variability was relatively predictable, there was a wider scale of unpredictability: 'a wet season might be late, or even worse, might fail altogether; a cyclone might tear great strips through the forests and inundate the coastal floodplains with a high tidal surge'.[14] Such events require that those who live there successfully understand variability and extremity and thus know how to survive.

The Jones and Meehan account is more like the Hoogenraad and Robertson account of central Australian seasons than it is like Davis's calendrical map which was produced in the same tropical area. Similarities between desert and tropics can be seen in the emphasis on big, relatively invariant forces — hot and cold in the desert, north-west and south-east winds in the coastal region. Within the interplay between these big forces is a great deal of variation and many local shifts, changes, and indicators. I term this system 'dynamic', in contrast to the relative stasis of the calendrical map. Variability is an integral characteristic of the dynamic seasonal system.

When imperialism takes over other knowledge systems, what is lost? Dick Kimber travelled with and learnt from people for whom conditions labelled by whitefellas as 'very changeable' were in fact 'entirely natural'.[15] This is a lovely statement of strong cultural difference around concepts of seasons, and it leads directly to the idea that the static map loses the dynamism and variability that characterise Indigenous meteorology from the Centre through to the Far North.

Calendars may thus promulgate epistemological error. The very act of mapping seasons suggests a standard or norm. Variation may thus come to be construed as deviation from the standard. This approach maps onto Aboriginal meteorology the problem that is inherent in so much of European people's attempts to settle Australia. It maps a desire for stability and predictability onto a continent characterised by extremes of variation and unpredictability, and implicitly defines as deviant any weather that fails to conform to these Eurocentric standards. It promotes exactly the error that through science and local knowledges we settlers are struggling to overcome.

There are so many studies that emphasise the variability and unpredictability of Australia's climate (the essays in the Webb volume are excellent examples) that there seems to be no good reason to promote predictability. Especially with the new research into the El Niño-Southern Oscillation (ENSO) and its effects on Australia's rainfall patterns, we are well into the domain of probabilities constructed around fundamental uncertainties.[16]

In addition to epistemological error, calendrical maps exclude large amounts of Indigenous knowledge because it is too localised, or too detailed, unpredictable or subtle to fit on a map. The calendrical map at Bowali Visitors' Centre, Kakadu National Park, shows how valiantly people have struggled to map dynamic relationships onto a static form.

We have to stop trying to fathom Australia according to ideas developed elsewhere and start to learn to live with it. But then we ask: what is 'it'?

Seasonal calendar, Bowali Visitor Centre, Kakadu National Park.

LIBBY ROBIN

Patchiness and precision

I will return to my own research to suggest that calendrical maps miss both the patchiness and the local precision of ethno-meteorology. Let us go back to hot and cold, dry and wet. The hot/cold contrast cuts across the wet/dry contrast and thus undermines the monolithic quality of the great sun/rain dyad. It is not so much that a dualism has been cut into more finely defined parts, but rather that it has been disrupted by the flux of the world. The key point here is patchiness — parts of each big division are scattered throughout the others. A part of each is in the other, an interpenetration that refuses dualism without eroding the integrity of either part. Cold can be seen to start when the rain starts to cool off the earth (that is, in the hottest time of the year), and so the rain of the hot time brings the conditions that will remove its necessity.

Local terminologies articulate the patchiness of rain: there is the very first rain in the hot time (*wuruwuru*), and there is the regular settled rain of the rainy time (*yipu*). There are light and dark rains, and there is the heavy rain. Then there is the cold weather rain that comes in the dry time (*kulwarang*). There is also a concept of the boundedness of rain. Hobbles Danayarri explained, 'first rain brings up the grass, last rain knocks im down'.

The cross-cutting of the contrasts wet/dry and hot/cold is attentive to the variability and patchiness of local climates. It subverts any form of complete dominance and allows for the interpenetration of elements: rain in the dry time, patches of cold air interrupting hot weather, patches of hot drifting into the cold time.

In sum, the wet/dry and hot/cold contrasts offer two underlying frames of reference. Wet/dry can be imagined as a digital system that alternates between the sun and the Rain(bow). When one is ascendant the other is not, and vice versa. Hot and cold (marked by constellations) can also be thought of in a digital mode, but they are also processes. Cold emerges from hot by the action of the rain, and hot emerges from cold by the action of the sun. Their actions are not uniformly continuous, but rather erupt in variable and unpredictable patches.

How, then, do Aboriginal people keep track of local seasonal changes? If it is all patchy and unpredictable, how do people even make decisions? Victoria River people organise their ecological knowledge through observation and experience of interactive local events. This is the patterned communicative domain of ecological time. The events of the world tell them what is happening almost day by day, and the information is abundantly rich.

Recall that the big picture is best thought of as interactive rather than prescriptive. Within this scheme, events are ordered by connections: sequence and co-occurrence (simultaneity) are of the first importance. When the march flies start biting, the crocodiles are laying their eggs. When the *Jangarla* tree (*Sesbania formosa*) flowers, the barramundi are biting. When the seed pods of the bauhinia tree (*Lysiphyllum cunninghamii*) turn dark red, the really hot weather is here. As Hoogenraad and Robertson state, this is a system that is very attentive to knowledge of water, and these hot weather indicators speak to ephemeral water.[17]

Living things communicate by their sounds, their smells, their actions, the stinging bite of the march fly, the sight of flowers floating in the water. They also communicate by their non-presence. Events that occur to the same rhythm require intervals of non-occurrence. There are times when things do not happen, and it is the not-happening that makes it possible for the happening to have meaning. Thus, for example, the march flies announce that the crocs are laying their eggs; their bites are only remarkable because for a long time they have not been around at all. Difference is regularly and predictably acknowledged, not erased, because many differences are of an order that makes a difference: they are information. The beat of simultaneity is made possible by intervals of non-presence or pause. Barbara Adam refers to this kind of time as 'nature's silent pulse', but of course in this system it is not silent.[18]

One of the questions I investigated was the extent of this system of local knowledge. I made a north–south survey that took me across two major ecological boundaries, and four different languages, including two language families. What interested me was: did the same system of signs occur across these boundaries? In summary form, the system holds across boundaries, and many of the same signs exist, but they index different events. So, for example, at Yarralin the march flies tell you that the crocodiles are laying their eggs. Slightly further south, but still on the Victoria River, at Daguragu, the flowering of a tree called *Jangarla* (*Sesbania formosa*) tells you that the crocodiles are laying their eggs. When it flowers, that's what it tells you. The message is the same, but the sign is different. Back at Yarralin, this same tree, when it flowers, tells you the barramundi are biting. Here the sign is the same, but the message is different. In Yarralin the bauhinia tree tells you that the really hot weather is here, while to the south-east in Mudbura country the same tree tells you about hot weather, but in a different way. Back in Daguragu, and further south again onto the great treeless plains of the desert fringe, one of the grevilleas, *Grevillea pyramidalis*, tells you that the hot weather is here when its seeds are cooked.

Old Jimmy of Daguragu contrasted *Ngumpin* (Aboriginal) knowledge with *Kartiya* (whitefella) calendars when he said: '*Ngumpin* going by trees, what the trees doing, and what pretty flower. When he get dry, we know October.' There is a highly detailed system of events that articulate moments within the broader interactive set. Some of these speak to the interaction between the rain and the sun whilst at the same time emerging from that interaction. Old Jimmy spoke of *Grevillea dimidiata* when he said '*Ngumpin* going by trees': '*walarrparr* tells you what time hot weather got to be, and cold weather got to finish'.

In sum, the rhythms of things that happen simultaneously are crucial codes for knowing what is happening in the world. Some of this knowledge is dispersed widely, but most is highly localised. Many of the signs of concurrence vary from place to place, so that one really only knows what is happening in the places where one has the knowledge of what concurrences exist and what they mean. The fact that this system is widespread ensures that people know there is a system here, and when they go beyond the bounds of their knowledge, they know that they are in the presence of a system they do not understand. The system is widespread, the content is local. All of it — system and content — is built from extremely long-term observations that have accommodated variability and patchiness and identified precise concurrences.

Conclusions

Aboriginal people have profound understandings of the ecology, including meteorology, of this continent. They have been here a very long time, they have achieved an enduring adaptation, and they have put their human powers of observation, analysis and communication to work to understand the parameters of the environments they inhabit and know as home. They are thus the possessors of a rich body of knowledge that is local, fine-grained, built up through long-term observations, and may be extremely sensitive to changing climatic conditions.

Without advocating appropriation in any shape or form, I am advocating dialogue. The fact is that Indigenous people hold in their songs, stories, myths, daily observations, and life practices the base-lines of knowledge that we all need. This is the knowledge that will enable all of us to know what changes are occurring and to make decisions about how we may continue to live our lives in this vulnerable continent. I go further to say that this is knowledge that holds human beings accountable in the world because they are connected into the world. What hubris is it to think that climate and climatic events have nothing, ultimately, to do with us?

Deborah Rose

Deborah Rose is a Senior Fellow in the Centre for Resource and Environmental Studies at the Australian National University. She is the author of *Country of the Heart* (co-authored with an Aboriginal clan), *Nourishing Terrains, Australian Aboriginal views of Landscape and Wilderness, Dingo Makes Us Human* (winner of the 1992–1993 Stanner Prize), and *Hidden Histories* (winner of the 1991 Jessie Litchfield Award). She has worked with Aboriginal claimants on land claims and in land disputes, and has worked with the Aboriginal Land Commissioner as his consulting anthropologist. Her work in both scholarly and practical arenas is focused on social and ecological justice. She is a Fellow of the Academy of Social Sciences in Australia. Her new book is *Reports from a Wild Country: Ethics for Decolonisation* (University of New South Wales Press, 2004).

4

Migrants and nomads

seasoning zoological knowledge in Australia[1]

Libby Robin

Living in a land like Australia with variable climates can be extremely challenging. In this chapter I review the migratory and nomadic strategies of birds in relation to seasons. I begin with shorebirds that spend most of their lives in Australia, but expend considerable energy flying north to Siberia for a few months, thereby escaping Australian conditions to achieve an annual breeding season. I then consider another strategy, adopted by the Banded Stilt, which is to sacrifice annual breeding seasons for sporadic, major breeding opportunities within Australia. European Australian zoologists have long recognised migration, but have been slow to recognise opportunism and aseasonal behaviour as adaptive strategies in birds. The chapter concludes with a brief reflection on how such avian strategies might have implications for the way a settler society lives with climatic variability.

Seasonal thinking and the migration model

In 1903, two Australian ornithologists made their way with considerable difficulty to the mouth of the Lena River in Siberia, 72 degrees north. They went to shoot some of their familiar birds in full breeding plumage. The urge to follow migratory birds to the ends of the earth had a long history — British fascination with birds whose movements followed and defined the seasons dated back to the Hampshire parson-naturalist, Gilbert White, whose famous book *The Natural History of Selborne*, was completed on the first day of 1788, the year that Arthur Phillip hoisted the flag at Port Jackson.[2]

In his letters to Thomas Pennant, White famously debated whether the swallows of spring migrated or went into hiding: 'I watched them narrowly this year', he wrote in November 1767,

> and saw them around till about *Michaelmas*, when they appeared no longer. Subsist they cannot openly among us, and yet elude the eyes of the inquisitive: and, as to their hiding, no man pretends to have found any of them in a torpid state in the winter. But with regard to their migration, what difficulties attend that supposition!

> that such feeble bad fliers (who the summer long never flit but from hedge to hedge) should be able to traverse vast seas and continents in order to enjoy milder seasons amidst the regions of *Africa*![3]

Birds and seasons seemingly go together. If you breed in Siberia, the season is short, sharp and circumscribed — but largely predictable. The annual long daylight of June–July triggers the breeding of mosquitoes, the tucker of the tundra which gives many waders and terns a reason to travel thousands of kilometres, year after year, to lay their eggs and raise their young in these food-rich, icy wastes.

The mosquitoes were wonderful for the birds, but terrible for the people. 'They settle on the gun barrel so thickly that you can't see the sight',[4] Ernie Trebilcock, the young Australian ornithologist complained in his diary. His more senior companion, Robert Hall, also found the boggy tundra heavy going:

> Our waggon sank to the axles, & once we had to un-harness the horses to get one of them out of a bog. In crossing a creek the waggon lurched & sent some of our luggage into the mud, & a little further on it upset most of it into the water. The result was that as soon as we reached some firm ground we pitched our mosquito tents & lit a large fire to dry our luggage ... we had to try to sleep without rugs that night. On all sides was swamp. Mosquitoes were in millions.[5]

Wagon and horses, Siberia 1903.

R. E. TREBILCOCK, AUSTRALIAN MANUSCRIPTS COLLECTION, STATE LIBRARY OF VICTORIA

They struggled on to finally reach a place '3,000 miles from the nearest railway station'.[6] They used river barges and then walked, to follow the migratory waders to their breeding grounds in June–July 1903. It took them five months to get there and the same to get home, via England. 'The birds would be [in Australia] before us', Hall commented wryly.[7]

Australian ornithologists, raised on British ideas like Gilbert White's, were fascinated by stories of migrants, who like themselves, had travelled from the far north to the far south. The question of where to find breeding grounds was of central interest to 'oologists', who practised the science of egg collection and classification.[8] The livelihoods of professional ornithologists depended on their knowledge of bird movements and breeding grounds. The early issues of *Emu*, the journal of the Royal Australasian Ornithologists' Union from 1901, focused on sharing this knowledge. Alexander Campbell, editor of *Emu*, needed egg collections and well-documented breeding grounds to write his major work, *Nests and Eggs of Australian Birds*.[9] Collectors were paid by wealthy international patrons to send them Australian eggs and birdskins. Oologists were acutely aware of birds known in Australia that bred elsewhere. Shorebirds, shearwaters and terns bred as far away as Siberia and Alaska, and collecting skins of these birds in breeding plumage paid for the Hall and Trebilcock Siberian venture of 1903. The remarkable thing was that there were, despite the careful observations of oologists all over the country, still some species of Australian birds whose breeding habits — and eggs — were not clearly identified and tracked. Such knowledge gaps were not due to a lack of effort, but rather because of an expectation that all birds, like the majority, would breed regularly and seasonally.

The exception: the Banded Stilt

Banded Stilts look and behave like many other waders, wandering in the shallows of salty lakes and periodically dipping and digging their long bills into the sands for food. They also often co-occur with other waders. On the 1896–1897 Calvert Expedition to the Kimberley, George Keartland recorded them in the company of their nearest relatives, 'Avocets and White-headed Stilts', near the junction of the Margaret and Fitzroy Rivers, although they are more commonly found much further south.[10] By the end of the nineteenth century, ornithologists recognised that breeding could occur over long periods in the tropics, but birds of the temperate south like Banded Stilts were still generally expected to be regular winter or spring breeders.[11]

It was surprising that there should be uncertainty about the eggs and breeding behaviour of a bird well known at Rottnest Island just near Perth. Many called it the 'Rottnest Snipe'. Banded Stilts were also well documented in the coastal flats of South Australia, hunting grounds of many of the nation's most enthusiastic and observant oologists. Ludwig Glauert and Clee Jenkins noted at a meeting of the Royal Society of Western Australia in October 1930, that a number of descriptions of the nest and eggs of the Banded Stilt had been recorded since E. P. Ramsay's first description of them in 1883. But in every case there had been controversy and the link to Banded Stilts was unproven; the eggs in most cases were those of the Black-winged Stilt.[12] The uncertainty about the location of rookeries fostered speculation — the birds clearly moved vast distances; could this be another transequatorial migrant?

Ornithologists knew that the plumage of many migratory waders changed at the time of breeding — in some cases taking on the rich brown and russet colours that provide camouflage in the tundra. The distinctive band that gives the Banded Stilt its vernacular name is its breeding colour — and it is russet like many tundra dwellers. The bird's scientific name *Cladorhynchus leucocephalus* focuses on its obvious white head rather than its optionally banded breast.[13] In non-breeding birds the band is reduced, mottled or perhaps even absent. The evidence of bands was not sufficient to deduce Australian breeding, as Siberian-breeding waders can develop their breeding plumage before leaving Australia.

'It seemed unlikely that the breeding places were outside Australia', wrote Glauert, palaeontologist and curator at the Western Australian Museum, who had conducted a full zoological survey of Rottnest Island, where the birds were usually plentiful, yet sometimes absent. 'Yet', he added, 'information concerning the "winter migration" was of the vaguest possible character'. He was intrigued that such a well-known bird could 'keep its [nesting] secret so successfully'.[14] Plumage was an indicator of breeding but was a cryptic clue to read, especially for Banded Stilts. In fact an expedition in 1995 documented some birds nesting before their plumage changed into breeding colour.[15]

The so-called 'winter migration' of the wet year of 1930 proved significant. In late July, Mrs B. E. Cannon, a Western Australian wheatbelt farmer from Kukerin, reported a huge rookery of breeding birds near her home at Lake Grace. This was very different to several earlier reports of 'Rottnest Snipe on this or that salt lake in the interior'. Mrs Cannon and her husband forwarded a blown egg and three photographs, asking for the name of the bird, and reporting 'thousands of them on a sandy spit running out into Lake Grace'. The museum immediately asked for more information, and on 4 September, the Cannons forwarded a parcel of twenty-three unblown eggs, one of which proved to contain a dead chick ready to hatch.[16] Glauert and his ornithological friend, the state entomologist, Clee Jenkins, realised they had the first solid evidence of where Banded Stilts breed, and at last they could confirm that they were indeed an Australian bird.[17] The find was of particular interest to Ivan Carnaby, an experienced oologist who was keeping close records of nesting in the Lake Grace area from 1928 to 1932. He commented that it was 'a place where very little [had] been done in the way of ornithology', but it was also 'typical of most of the inland portions of Western Australia, having chains of salt lakes and flats, with highlands of sand plain, ironstone country, granite formations etc'.[18] It was he that reported the sad end of Mrs Cannon's huge rookery on the peninsula near her home: it 'was despoiled (before half the eggs were hatched) — perhaps by foxes'.[19] Carnaby, an observant bush naturalist, was to continue to contribute important specimens and observations to arid zone bird science throughout the 1950s and 1960s.[20]

Before that wet year of 1930 was out, the idea that Banded Stilts bred in 'winter' was challenged by finds at Lake Callabonna, north of Lake Frome in South Australia's north-east. The lake, famous for fossil Diprotodon finds in 1892, provided a different sort of first for J. Neil McGilp and Matt Morgan, long-time stalwarts of the nation's oldest bird group, the South Australian Ornithological Association.[21] In late December, McGilp and Morgan heard reports from Mr R. McKay, manager of Moolawatana station, of a rookery of nesting birds on a previously unvisited island in the lake. On 11 January 1931, they followed his directions and waded out a tedious 'three-quarters of a mile' through 'two feet of water and six inches of mud'. To their amazement and delight their efforts were rewarded with 'a densely packed mass of Banded Stilts' — about 54,000 of them — which took to the air 'with a whirr of wings, raising quite a dust as they did so, and revealing thousands of eggs on the bare ground'.[22] There was no protection from the merciless heat (recorded at 104°F in the shade (40°C)) or the hot wind, the only vegetation being 'a few samphire bushes'. As McGilp and Morgan commented:

> It seems somewhat extraordinary that the second discovery should have so soon followed the first, after the nesting of these birds remaining a mystery for so many years.[23]

The Banded Stilt breeds 'only in the arid inland at large shallow ephemeral saltlakes inundated by rain or floodwaters',[24] in which brine shrimp, its main food, irrupts. The presence of the brine shrimp is crucial to the breeding event; not the time of year, nor whether it is hot or cooler. McGilp and Morgan noted the 'abundance of food' in Lake Callabonna in January 1931 with the water 'swarming with small crustaceans of the genus *Apus*'.[25] Brine shrimps were also 'present in millions' in Lake Grace in September 1945 when the Stilts nested there again, the first time since July–August 1930. The rainfall in Perth in 1945 was a record 52.67 inches (over 150 per cent of average rainfall). Ivan Carnaby tested the water at Lake Grace and found it to be 'salter than sea-water'.[26] There was some evidence from Aboriginal people that a major breeding event had occurred at Lake Callabonna in 1918, the last time the lake had filled before 1930.[27] Twice in a year — or not at all for twelve or fifteen years. Such are the vagaries of Banded Stilts' breeding events. Neither its place nor time of breeding is predictable, though there is some evidence that the two events of 1930 actually affected different populations of birds — one in southern Western Australia, and one in eastern South Australia.[28]

Opportunism and nomadism — another model?

The Banded Stilt discoveries were part of international attempts in the 1920s and 1930s to understand the physiology of irregular breeding — how local conditions such as rainfall might 'trigger' physiological responses. John Baker, working in New Guinea and the New Hebrides in the 1930s, posed the question of whether tropical birds were capable of year-round reproduction. His fieldworkers, Tom Harrisson, later famous in Sarawak, and the young Jock Marshall, also learned to question the assumption that changing daylength is *the* trigger for breeding.[29] Marshall, later working with Western Australian ornithologist, Dominic Serventy, extended Baker's questions about tropical latitudes into the archetypal Australian landscape, the arid zone.

Deserts (broadly speaking) cover about 28 per cent of the earth's land surface, if they are defined by heat as well as aridity, which was the practice for much of the twentieth century. Although most deserts are in the populous Northern Hemisphere, they are the least populated places on earth in the modern era. Generally this puts them far away from the centres of western science. The struggle to come to terms with the extensive Australian arid zone and its uncertain seasons was late in coming. The post-war era of the 1940s and 1950s saw Australians looking again and differently at the 70 per cent of the country not governed by regular seasonal rain.

The term 'desert' is problematic and difficult to define with precision and independence. As Dom Serventy declared in despair in 1971: 'A desert occurs wherever it is said to occur. A definition by acclamation!'[30] Deserts are typically defined in human terms as places where 'dryness and heat are so excessive that normal agriculture is impossible in average years'.[31] More recently, cold deserts have also been included.[32] They, too, exclude agriculture. Scientists now prefer the term 'arid zone'. This is defined by balancing the effects of precipitation and evaporation and declaring it to be a region of low rainfall in higher latitudes and higher rainfall in tropical latitudes. But it is still an unsatisfactory definition. The arid and semi-arid zone embraces not just low rainfall and high evaporation areas, but also those with great variability and uncertainty of season. The Arid Zone Research Centre in Alice Springs defines the zone in Australia by rainfall, evaporation and also the quality and usefulness of its vegetation, something that is not necessarily annually or seasonally dictated. Arid and semi-arid lands are:

Banded Stilt colony.

CLIVE MINTON

> bound by average annual rainfalls of about 250 mm in the south but up to 800 mm in the north and about 500 mm in the east. Together with sub-tropical regions and the mountain high plains, they form the rangelands, where rainfall is too low or unpredictable or where terrain is too inhospitable for sustainable cropping or timber harvesting.[33]

The problem is not so much the dryness, but not knowing when, or where, the next rain will fall. Averages can be very misleading. It is easier to grow wheat in the Western Australian wheatbelt on ten inches (25 to 30 mm) average rainfall than in similar poor soils in north-western Victoria on sixteen inches average rainfall, because in the Western Australian wheatbelt the time and quantity of rainfall are more predictable.[34] In parts of the Victorian Wimmera and mallee, some seasons have less than 30 per cent average rainfall. Further out in the rangelands of the semi-arid zone, beyond wheat, there are years where no rain falls at all. What are the breeding patterns for birds and other animals in a land with such irregular environmental cues?

What concepts do we need to understand places of variable and erratic dryness? Mark Stafford Smith and Steve Morton, in an important paper on the ecology of arid Australia and its management implications, observe that all arid regions have spatially and temporally variable rainfall. Australia's particular problem is that its temporal variability is higher than elsewhere. This means that most years have less than average rainfall, with just a few years of very high rainfall skewing that average.[35] Small isolated rainfall events may not stimulate plant growth much, while big rains, caused by cyclonic and monsoonal depressions, can have major and widespread effects on the biota aided and abetted by the prevailing flatness of inland Australia.

The temporal variability is complicated by the bimodal possibilities of rain in the Australian arid zone, as Henry Nix has argued: 'Much of the arid zone is only moderately seasonal' in terms of rainfall.[36] Rain does not fall often, but it may fall in the summer (an extension of the northern Australian monsoon rains beyond their normal range) or in winter (a northerly extension of the normal southerly driven rains that extend across the southern parts of the continent and Tasmania). Such a pattern, coupled with the long-term history of the region drying out (over geological time) makes for distinctive evolutionary patterns of both flora and fauna.[37] Nix summarises his plant data as showing that the northern parts typically have 'megatherm' plants whose photosynthesis is best at 30–33°C, and can tolerate up to 46°C; and the southerly extreme tends towards 'mesotherm', plants whose optimal photosynthesis is 19–22°C, going up to 33°C. In between, however, there is 'wide overlap'. In glacial times, different 'microtherm' (frost tolerant) flora would also have been prominent. This overlapping of megatherm/mesotherm plant zones with low, unpredictable rainfall supports unusual fauna.

It is this complexity of temporal variability and bimodal rainfall patterns that supports a double breeding season for birds widespread in the interior of Australia.[38] Thus we effectively have two 'springs' — after northern summer rain or after southern winter rain — and can discern a weak tendency to 'spring' breeding — but the location is entirely dictated by where the rain has fallen.

The Australian arid zone ecosystems are dominated by termites and other invertebrate and reptilian predators, not warm-blooded (homeo-thermic) birds and mammals who are, like ourselves, physiologically poorly adapted to cope with climatic extremes and variability.[39] In these parts, rain does not have to fall from the sky — water can run through channels, lakes and normally dry river beds. Nearly half the birds of the desert depend on the availability of free water; but because nutrients and surface water move together, it is often food, rather than water, that creates breeding opportunities for arid zone animals.[40] The brine shrimps, for example, are crucial to Banded Stilts. Water is a prerequisite for the shrimps, but in their absence birds will abandon their rookeries even if the lakes still have water in them.[41]

Internationally speaking, most arid zone birds still breed seasonally. The observant Reg Moreau, long-time editor of the British journal *Ibis*, who had access to data from all over the former Empire, discerned an anomaly in the Australian arid zone. In a footnote in 1950 he declared that 'no African bird community has become so thoroughly adapted to such a climate as have the birds of north-western Australia'.[42]

Following the Australian Museum's expedition to central and north-western Australia in the grip of the 1952 drought, Allen Keast and Jock Marshall adopted an integrative view — they melded studies of breeding biology, reproductive physiology and environment and began to piece together a story for the exceptional birds of the Australian arid zone, expressing the view that drought and rainfall were key influences on breeding amongst desert birds.[43] In 1959, Marshall declared in his typically bombastic way that 'a rigid adherence to any particular phase of periodicity would be disastrous [for arid zone birds]: such birds breed irrespective of the cycle of the sun'.[44]

In the 1950s, a German physiologist and systematist, Klaus Immelmann, studying arid zone birds in Africa while working in the dry Kalahari and extremely dry Namib, found regular seasonal patterns of breeding. Inspired by Moreau's remarks and the work of Marshall, Serventy and Keast, the young but well-qualified Immelmann came to Australia in 1959–1960, funded by the German Federal Research Organisation, to compare the exceptional Australian arid zone with his earlier work in Africa.[45] He was one of the first ornithologists to spend a wet season in north-western Australia, and was witness to the breeding response of Zebra Finches *Taeniopygia guttata* to drought-breaking rains in 1959. He publicised the 'opportunism' that he had witnessed and this changed much international thinking on the matter. His view was that it was the sight of rain that stimulated the breeding response in finches and in Black-faced Woodswallows *Artamus cinereus*.[46]

Both Marshall and Immelmann died relatively young with their life's work in arid zone avian biology incomplete.[47] Other reproductive physiologists turned their attention again to wild birds living under arid zone conditions only as late as the 1990s, and returned to the question of birds with particular local knowledge of environmental resources and their limitations. Richard Zann, for example, a great admirer of Immelmann, modified his theory about the Zebra Finches' breeding response to the sight of rain to include the sight of new grass shoots.[48] Lee Astheimer and William Buttemer looked at an arid zone population of White-plumed Honeyeaters, *Lichenostomus pencillatus*, one of Australia's old endemic passerine species. They found them to be like Zebra Finches in that they could 'break the rules' of cyclic breeding and nesting opportunistically. Females could lay eggs in any month of the year, and males were always 'at the ready', retaining enlarged testes all year round.[49] As Richard Schodde summarised for a large variety of arid zone birds: 'of all behavioural traits suited to deserts, nomadism is by far the most important'.[50]

Waterbirds and shorebirds in Australia are apparently physiologically similar to those of the Northern Hemisphere, but their breeding behaviour has been very significantly modified by environmental conditions. Whilst birds of the north are generally characterised by regular migrations — they breed in the northern spring and early summer, then winter in warmer climes further south — many of Australia's endemic birds move unpredictably. Day length and seasonal changes in temperature are not the only triggers for breeding behaviour in many nomadic breeding birds of the Australian arid zone. Nomadism is a response to our erratic climate for all sorts of birds and other animals that depend on water, insects, vegetation and other animals as food sources. The paradoxical and ephemeral wetlands of the arid zone provide alternative possibilities for many birds that also breed in salt lakes nearer the coast. Examples include pelicans, waders, cormorants and freckled ducks. But only one bird depends entirely on the ephemeral wetlands of the arid zone for its breeding habitat, and that is the Banded Stilt.

The Banded Stilt is another sort of keystone species — not a bird representative of a range of like birds, but an extreme case of a bird whose physiological adaptations emerge from older climatic patterns, that lives now in the relict and ephemeral wetlands of a formerly better-watered interior. Its total dependence on arid zone wetlands for breeding grounds alerts us to the importance of these 'last resorts'.

Other birds may need the arid zone too — and although they may not breed there exclusively, opportunism may be an important part of their breeding behaviour and their population ecology, as recent work by Julian Reid and Roger Jaensch suggests.[51]

As sources of water outside the arid zone are increasingly regulated, the need for these wild and unregulated — but ephemeral — wetland resources may become more critical, even for species commonly regarded as unthreatened. Birds as common as pelicans, cormorants and swans have disappeared in living memory from Perth and its suburbs as the historically named Swan River is constantly reclaimed for real estate and roads.[52] Dom Serventy would never have realised that his passionate work on arid zone birds, undertaken as a hobby outside his work at CSIRO, would provide, within a few decades, critical insights about the fish-eating birds of his suburban morning walks. Where can a fish-eating bird go and thrive if the changing river can no longer support fish? The answer is 'somewhere outback'.

Colonial waterbirds and waders may need the erratic, unexpected, opportunistic moments afforded by arid zone wetlands to prop up populations — even where they sometimes breed closer to the coast and in more richly watered areas. Because they can turn up almost anywhere there is a suitable water source, and ephemeral water sources can move widely in the flat interior, it is difficult to plan for breeding events and to save or reserve 'key areas'. In the scientists' jargon, the birds are 'poorly differentiated biogeographically', but this does not mean that they are not highly dependent on such areas as the irregularly-watered but biodiverse Channel Country of south-western Queensland. Following the large Lake Eyre basin floods of 2000 and 2001, Julian Reid found enormous levels of successful breeding among a wide range of waterbirds (for example, pelicans, cormorants, herons, ibis, ducks, shorebirds and terns) in the Channel Country and in the adjacent Coongie Lakes region in north-eastern South Australia. 'There are records of breeding in this bioregion by almost all species of waterbird that breed in Australia and occupy central Australia on a non-vagrant basis', Reid noted. In fact, 'the temporary wetlands of arid Australia may be preferred breeding habitat for many species' in the face of changed conditions in their more traditional breeding grounds.[53] The idea of 'preferred habitat' may be of conservation significance, and it may be important not to wait till it's the 'last habitat' in view of its irregular availability.

In 1968, Harry Frith wrote a document for the newly established Australian Conservation Foundation's *Viewpoint* series on conservation priorities for waterbirds. In it he claimed the arid zone 'occupies 1 million square miles, more than half the continent and is virtually useless to waterfowl as no permanent habitat exists there'.[54] In the same year the group drawing up priorities for a proposed biological survey of Australia suggested that marine biology might be among the fields which could be excluded, as 'relatively less urgent than terrestrial biology'.[55] As we learn more of coral bleaching and other threats to marine resources — including those by some of the same agricultural chemicals that upset the terrestrial biota — this advice to government looks poor. It was certainly limited by the knowledge of its day.

The danger of dealing first — and therefore often only — in 'priority areas' is that omissions can prove fatal. Conservation managers and governments have to maintain lateral vision, despite the pressure to research efficiently. We cannot fund everything, but we cannot always simply go with the majority view all the time — other voices, the beating of other drums, may become significant in the medium and longer term. We must make space for alternative conceptual models to survive, despite the pressure to go with the majority and the latest findings. 'Sustainability' depends not just on the views of the here and now, but also on our ability to hear, and to some extent accommodate, the stories from other places and times. 'Difficult' research areas — like marine science and the erratic, unseasonal science of the arid zone — should not be unsupported because they are slow in producing results. Their importance may change with time. Priorities are to some extent annually (or triennially) dictated. Environmental managers concerned about long-term futures have to resist being annual-fashion slaves.

Nomads and their cultural baggage

The idea of nomadism was applied to people many years before the bird stories. The observation that some Aboriginal people moved around seasonally was applied to all Aboriginal groups, some of whom were almost totally sedentary. It was a tool of empire from the earliest days of European occupancy, denying Indigenous land ownership and supporting the myth of *terra nullius*. In fact, no Aboriginal groups were truly 'nomadic', in the traditional sense of moving to maintain conditions of pasture for animal herds.[56] Aboriginal movements were motivated more often (like the Europeans' own) by trade, ceremony and the availability of resources. But Aboriginal people were 'other' and so, traditionally, were nomads.

In his 1841 *Essay on History*, Ralph Waldo Emerson declared that all history was encompassed in the individual. Just as the acorn is the key to a thousand forests, so history is the sum of the universal human mind. But the nomad was on the wrong side of the Western history that Emerson saw as universalising:

> In the early history of Asia and Africa, Nomadism and Agriculture are the two antagonistic facts. The geography of Asia and Africa necessitated a nomadic life. But the nomads were the terror of all those to whom the soil or the advantages of a market had induced to build towns. Agriculture therefore was a religious injunction because of the perils of the state from nomadism ...[57]

But although the 'we' of this history seemed to work in praise of agriculture and the towns, Emerson dispelled this quickly. In the next sentence he contrasted the nomadism of Asia and Africa with that of America and Europe, the latter being 'of trade and curiosity'. He also praised 'intellectual nomadism' as 'the faculty of objectiveness or of eyes which everywhere feed themselves'. 'Who hath such eyes', he declared, 'everywhere falls into easy relations with his fellow-men ... His house is a wagon; he roams through all latitudes as easily as a Calmuc'.[58]

This was clever, but required significant logical slippage. The nomadism of Asia that sought new pasturage from month to month, or that of Africa, which escaped the gad fly that drove the cattle mad, were locally specific conditions, determined by environmental forces. But the nomadism of the West was driven by its own universalising culture. Only cultural nomadism, not locally-informed, environmentally-aware nomadism, could live on a wagon that could go anywhere. It both subsumed and transcended agriculture.

Thinking like a Banded Stilt

Irregular, uncertain and sporadic seasons are at the heart of the 'otherness' of the Australian inland for the well-seasoned, European eye. It is natural to focus first on the natural history that confirms our own place, and endorses the arduous travel that took us (or our forebears) to the 'end of empire'. Migrants are 'us' in the neo-Europes; nomads are 'them'.

Cultural preconceptions have a tendency to shape the focus of science and understandings of fauna and place in subtle ways. The stories of Siberian migrants make natural — even unremarkable — the travels of the European to the antipodes. Aldo Leopold famously urged forest managers in the United States to break out of their cultural preconceptions and take the perspective of natural elements, to 'think like a mountain'.[59] This echoed his compatriot Emerson who one hundred years earlier 'held our actual knowledge cheap', because we know so little 'sympathetically or morally' of the life of non-human nature, such as rats, lizards, fungus or lichen. The Indian and unschooled farmer's boy, he declared, 'stand nearer to the light by which nature is to be read than the dissector or the antiquary'.[60] He urged 'an ethical reformation', and declared 'the path of science and of letters is not the way into nature, but from it rather'.[61]

If we are to understand the arid zone and its seasons and seasonlessness, we might start with Arrernte knowledge. Take, for example, the information portrayed on a board at the Alice Springs Desert Park.[62] Here we find an only partly annual 'calendar', with a regular hot and cold season and another, irregular season — simply called 'after rain'. Rain is not seasonally distributed, but when it occurs, the 'after rain' season happens. Such knowledge is not easy to plan around.

Arid zone biologists, however, are increasingly infiltrating the concept of 'variability' into discussions of rangelands management, in an echo of uncertainty and chaos in the physics of the new millennium. What we have in Australia is a nomadism that is not simply temperature-driven, like in the Northern Hemisphere, but is a complex function of temperature, water, vegetation and other life-sustaining resources.[63] Banded Stilts are one of a number of species that can teach us much about living with uncertainty, to take opportunities as they arise, but not to expect regularity.

The lack of regularity in their cycle challenges all the fiscal models built around annual thinking. The global economy is annual. European and North American economies, based in seasonal agricultural regularity, shape and control the world's fiscal thinking and practice. There is no economic model that follows the patterns of the Banded Stilt. It is much easier to make international treaties to protect migratory birds on regular annual flights along international 'flyways' than to manage rangelands for wildlife within a single nation for unpredictable, non-annual seasons.[64]

The Banded Stilt suggests that we may need to rethink Australia's place in the world from the inside out. Historical geographer Joe Powell has described the populated, seaward-looking cities of the nation as 'the restive fringe'. From the time of Federation, the people of the imperial cities looked outward and overseas for models of settlement, planning and economy.[65] All major cities in Australia have seasons (tropical or temperate) and are relatively remote from the arid zone. The people of the cities predominantly looked back to where they came from — and to other places where the British ventured. The 'Anglo World' described by Thomas Dunlap in his book *Nature and the English Diaspora* was dominated not so much by Britain itself, as by its imperial legacy — its Gilbert White models with echoes of the hollow lanes of Hampshire in many places where nature is different.[66]

Different models exist in the Australian outback, but we have been slow to understand them and late to argue for their significance in national or global discourse. Historically we have suppressed desert knowledge — it was regarded as unpatriotic to mention the Australian desert. The Banded Stilt throws out a challenge with its nomadic and opportunistic perspective. How might we accommodate the needs of a climatically variable, arid land in the overwhelmingly seasonal worldview that is one of the legacies of empire?

Libby Robin

Libby Robin is a historian of science and environmental ideas. She is author of *Defending the Little Desert: the Rise of Ecological Consciousness in Australia* (Melbourne University Press 1998) and co-editor (with Tom Griffiths) of *Ecology and Empire: Environmental History of Settler Societies* (1997). Her most recent book, *The Flight of the Emu: A Hundred Years of Australian Ornithology* (Melbourne University Press 2001) won the inaugural Victorian Premier's Award for Science Writing in 2003. She is a Fellow at the Centre for Resource and Environmental Studies, Australian National University, Canberra.

This single furrow horse-drawn plough was used by wheat breeder William Farrer at his experimental farm near Canberra between 1885 and 1906. Often referred to as the 'father' of the Australian wheat industry, Farrer successfully bred new varieties that thrived in Australia's challenging climate.

English wheat varieties imported by colonists in the nineteenth century often failed in summer droughts or succumbed to diseases such as stem rust. These strains matured late in the season and were best suited to wetter northern hemisphere summers. Towards the end of the century, work began on breeding wheat varieties that could withstand the drier conditions of the Australian mainland.

Farrer developed a range of wheat varieties that matured quickly with well-developed heads on shorter stems. Early maturity of these strains helped them avoid the rust fungus, which affected crops that matured later in the season. In 1901, Farrer decided to name a variety with a particularly high yield 'Federation Wheat', in honour of the recent union of Australian states. Federation Wheat became the most popular wheat variety grown across the country between 1910 and 1925 until new varieties overtook it in the 1930s. It was also successfully introduced in India and the United States, where it was in turn crossed with other varieties.

Single-furrow horsedrawn 'colonial' plough, used by William Farrer at Lambrigg, near Canberra, about 1900, donated by B. Tong

NATIONAL MUSEUM OF AUSTRALIA
PHOTOGRAPH BY GEORGE SERRAS

The Possibility of Rain

Wheat samples of the past and future: (*left*) white lamas, historical wheat variety grown by CSIRO Plant Industry, 2000; (*middle*) Federation wheat, historical wheat variety grown by CSIRO Plant Industry, 2000; (*right*) Line B73–6.1, transgenic wheat variety developed and grown by CSIRO Plant Industry, 2000

NATIONAL MUSEUM OF AUSTRALIA
PHOTOGRAPH BY GEORGE SERRAS

5

Goyder's Line
the unreliable history of the line of reliable rainfall

Janis Sheldrick

South Australia is notoriously 'the driest state of the driest continent', and the only state commonly identified with the continent's broad climatic norms rather than with aspects of its diversity. But along with aridity, Australia's climate is characterised by widespread high rainfall variability and, as the historian of Aboriginal Australia, John Mulvaney, has put it, it is 'seasonal reliability or variability of rainfall, and not its average, which is the true gauge to the ecological situation'.[1]

While investigating the life of George Woodroffe Goyder, the Surveyor-General of South Australia from 1861 to 1894, I came to understand that by 1865, after just over fourteen years in the colony, he had not only grasped the reality and impact of unpredictable rainfall, but had put his knowledge to use mapping the border at which it became too great to sustain European-style agriculture.[2] The border became known as 'Goyder's Line', or 'Goyder's Line of Rainfall', a feature of maps of the state for over a century. The line has been the subject not only of contention in its role as an agricultural limit, but also of conjecture as to what it is and how it was drawn. The mystery of the line's meaning, though, turns out to be itself the curiosity, because George Goyder had plainly stated what he intended to map — reliable rainfall — before the line was drawn, and he went on warning about unreliable rainfall for the next thirty years. The history of the line demonstrates that he was not truly heard, and the deafness of his contemporaries and their successors inhabits the gap between the experience of weather — droughts and flooding rains in this case — and ideas about climate.

Goyder was a person whose imagination and energies — and he was famously energetic — were stimulated by the environment of the new continent. In South Australia he expanded in a way which would not have been possible in England. He had arrived in Sydney in 1848 as a young engineer from the industrial north, settled in Adelaide in 1851, and joined the Land Office in 1853. It was not long before he was effectively second in command of the Survey Department, and by the end of 1860, he had been chosen to take over after Surveyor-General Freeling's return to England. In the decades that followed he expanded the Survey Department into a broad-ranging administrative empire, and reached beyond its bounds, initiating activity, exercising authority and creating administrative structures in areas such as forestry, water resources and the drainage of the south-east. For over two decades he was the South Australian government's 'universal genius', as one commentator at the time put it, and selected and surveyed the site of Darwin, finally establishing permanent colonial settlement there after a series of failed attempts.[3] The line was only a part of his enormous output of work, but it is a central, important and intriguing part.

Goyder's Line

Since Goyder's definition of his line escaped his contemporaries, the history of the line has been shaped by incomprehension and confusion. It is generally understood that the line was produced in 1865, during the great drought of the mid-1860s, as an administrative tool to enable the provision of drought relief to the holders of pastoral leases in badly affected areas. It was not promoted to the role in which it would become locally famous until the great wheat colonisation of the 1870s, later the subject of the American scholar Donald Meinig's classic work of historical geography, *On the Margins of the Good Earth*.

The purchase of agricultural land on credit, in designated Agricultural Areas, had been introduced at the end of the 1860s. When these areas proved inadequate, the line was pressed into service, in August 1872, as the northern limit of the country to which the credit provisions applied. The 'earth hunger' of the settlers was not sated by this, and the volume and intensity of the pressure to repeal the limit was such that as Goyder himself later put it: 'no Ministry could remain in office that declined to bring the land onto the market'.[4] The line's short official career as a limit was ended in November 1874, just over two years after it had begun. As each area beyond the line was made available for selection, Goyder warned in his description of the country that the rainfall within the proposed survey area was doubtful, and the results of farming likely to be precarious. It was to no avail:

> when survey and opening the land followed it was taken up more readily than before. The favourable seasons appeared to justify such action and to throw doubt upon the conclusions of previous experience. 'The rain follows the plough' became the cry, and the demand still further and more urgently increased ...[5]

In what has been described as 'the most outstanding episode in the history of South Australia before the twentieth century', thousands of people poured into the north, and vast areas were transformed into wheat lands.[6] But as Goyder grimly recalled, 'the seasons changed at length ... failure ensued in the outside hundreds, selection all but stopped ... the surveys were discontinued ... ' The change occurred at the end of the 1870s, and throughout the 1880s: 'the seasons continued bad and failure followed failure ... '[7] While the 'Rainmaker-General' and his imaginary line had earlier been the butt of jokes, in 1882 the parliament heard an excerpt from a letter in which a disconsolate farmer observed that Goyder's ghost seemed to hover about (though Goyder was far from dead).[8] The retreat from the north that followed over the next decades left the degraded country littered with deserted buildings.

Goyder's vindication impressed itself on his contemporaries, but not in such a way as to stop the story repeating itself less dramatically, especially as the mallee was cleared in the east. When settlement and land use patterns eventually stabilised, the line ceased to be a focus of contention, although it remained a part of South Australian culture, and the 'inside' and 'outside' country it created became part of the language of the landscape. It is still a point of reference in television weather forecasts.

Although the line had been produced during a major drought, the idea that it was more than just a drought border was abroad during Goyder's lifetime, and persisted beyond it. Many suspected a connection with changes in the vegetation, but the most commonly proffered account came to be that the line was a clever approximation of an isohyet of average annual rainfall — an estimation of aridity rather than variability. That understanding was taught to generations of South Australian schoolchildren. It was so widely accepted that in the heading of an article in 1963 celebrating Goyder's other achievements, he was identified as 'Surveyor Goyder of the Rainfall Line ... the man who drew a line to depict average rainfall'.[9]

Navigating the inland

Attempting to provide a history of Goyder's Line as he understood it inevitably raises the questions of how he came to be able to recognise rainfall variability as an enduring and ecologically critical aspect of climate, how he was able to map his insight, and what had motivated him. He was not a scientist, and had no body of formal observations to work with, other than the Adelaide records of his good friend Charles Todd (identified by Neville Nicholls in Chapter 2 as one of the first to recognise a 'teleconnection' of El Niño events). The answer to the first question seems clear enough: Goyder's focus was a response to his particular experience of the climate of the inland.

In 1857, after three years as Assistant Surveyor-General, Goyder had been given an opportunity to do a little exploring in the country between the end of the Flinders Ranges and Lake Torrens, the gigantic horseshoe lake that was believed to form an unbroken and impassable barrier of dazzling white salt and mirage. Against all expectation, Goyder saw 'vegetation of the most luxurious kind' covering the valley floor as they travelled toward the lake along the MacDonnell Creek, though as they moved out onto the plain he noted 'the timber in the creeks changing from lofty gums to bastard peppermint, which was rapidly assuming a more stunted appearance'.[10] Eventually he and his two companions reached the lake at what is now Lake Blanche. It turned out to be full of fresh water, which extended to the horizon. After investigating its treeless shores, he concluded that the waters were permanent.

This amazing news triggered an outflow of breathless editorials, geographical speculation, and a pastoral land rush from the eastern colonies. Captain Frances Cadell (who had explored the Amazon and the Murray) offered his services for a steamboat navigation of the unknown waters.[11] But the party which struggled back through the Flinders Ranges, with a galvanised iron boat nearly six metres long and a smaller, folding iron punt, was led by Surveyor-General Freeling. At the site of the vast lake, they floundered in shallow water and mud amidst desolate surrounds. Freeling rebuked Goyder for not having paid 'closer attention to the general character of the country', and explained that his hasty and over-enthusiastic young assistant had seen floodwaters, and been beguiled by mirage into mistaking their extent.[12]

Although there was considerable anger at the disappointment, the mistake was dismissed as that of a well-meaning novice in exploration. Unfortunately, because Goyder was thought to be mistaken about the extent of the waters as well as their permanence, the real point of his experience was lost. It appears that only one writer, the Reverend Julian Tenison-Woods, attended to the implications of a large body of floodwater in the centre of the continent, even while criticising Goyder for exaggerating its extent.[13] Goyder and his companions seem to have been the first colonists to witness extensive flooding in the inland, and the dramatic transformation of the landscape that follows, but it was not until thirty years later that the veracity of Goyder's basic observation of the extent of the water was borne out, when widespread flooding occurred again. His partial vindication was made public the following year in Andrew Garran's *Picturesque Atlas of Australasia*, published in 1888, but it was too late to be newsworthy, and the story of his encounter was never reinterpreted.

In 1859 Goyder returned to the north, during the worst drought the colonists had so far experienced, and led a large expedition into the country south of Lake Eyre. The parties were camped in the region for eleven months and triangulated a huge area. While they were at work, the drought broke. Once again, Goyder saw volumes of water and the country transformed, but he was wary of reporting it and waited until he could return to the lake and make a second examination. In a venture highly appropriate for one who would be twitted as the discoverer of the Inland Sea, he took two companions and boated down the Margaret Creek to the edge of the lake — which was drying — in a coracle constructed from bullock hide.[14] It was probably during this expedition that he saw Aborigines breaking camp to move toward distant storms, a practice which brought him to the realisation that this was, as he put it, 'nomadic country'.[15] Goyder concluded that pastoralists could occupy this country if they developed similar opportunistic strategies and moved with the water. When he was recalled to take up the position of Surveyor-General, Goyder returned to Adelaide with an insight into the climate of the inland that was unique among colonists, certainly among those in power.

Learning to see

The route to mapping his insight was longer and less brutally direct. Goyder was relentlessly curious, and was described as a keen observer, whom nothing escaped.[16] Even before his expedition to Lake Eyre South, his knowledge of the country was perceived to be 'intimate'.[17] Since these powers of observation were highly schooled and this 'intimate' knowledge hard won, the course of their acquisition can be traced, beginning from the point at which Goyder chose to travel overland from Melbourne to Adelaide. Once there he met fellow Swedenborgian W. S. Chauncy, an experienced colonial surveyor, who was interested enough in the new colony and its natural environment to produce a book describing it in detail. With Chauncy as best man, Goyder married Frances Smith, whose brother and father, both in the colony, were nurserymen. It was a liaison that fostered an interest in trees and plants, and a growing botanical knowledge. From 1854, when triangulation began again toward the head of Gulf St Vincent, Goyder was in the field, and in 1855 he became First Assistant Surveyor-General, responsible for survey parties. This kept him travelling constantly around the edges of the survey, which, in the ordered colony of South Australia, radiated contiguously from Adelaide. (Country beyond the surveyed areas, or 'hundreds', could be leased for pastoralism, until required. Much pastoral country was triangulated well ahead of the survey to enable the boundaries of runs to be fixed.) Even before this appointment, Goyder had suggested changes to procedure to ensure that the country was properly examined before work began.

By early 1859 he was seen as the inevitable choice of guide to conduct the visiting geologist, Alfred Selwyn, on a tour of inspection of the colony that would extend into the Flinders Ranges. In the course of the journey he seems to have learnt how to make and record systematic geological observations, and immediately included them in the survey routine of the Lake Eyre South expedition. This involved making observations about topography and vegetation along lines between trigonometric points, usually on hill tops, and Goyder proudly developed it as a fast and trustworthy method of map-making. The expedition to Lake Eyre South was as far inland as Goyder was to travel. In 1862 he made his furthest journey to the west into the country north of Fowlers Bay, and the following year he examined the south-east as a member of a government tour of inspection.

All of his skills and experience were called upon when he undertook the revaluation of the pastoral leases in 1864. As he criss-crossed over eighty runs, some of them huge, Goyder methodically mapped and studied each lease, recording observations of vegetation at estimated intervals along traverses. The first runs visited were the large holdings in what is now the prime agricultural country of the mid-north. The second batch were runs in the northern country just beyond, which was already beginning to dry as the long drought of the mid-1860s set in. The valuation reports on some of these runs contain very brief observations about areas of reliable and unreliable rainfall.[18] Goyder rode for thousands of miles — he estimated a seemingly impossible 30,000 miles — notebook in hand, studying the changes in the communities of trees, bushes and grasses, and recording the availability of water. The documentary evidence consists only of raw observations or finished reports, with no obvious traces of the intermediate steps by which Goyder's understanding developed, but presumably he had come to his own conclusions about the reliability of the rainfall and what he saw about him. His successor, William Strawbridge, who had served under him from the time the line was drawn, confirmed his interest in the location of saltbush, but saltbush was only one indicator among many.[19]

Mapping the line

While studying the country, Goyder had also had opportunity to reflect on a slowly but inexorably approaching crisis. It was obvious that the amount of land suitable for agriculture was limited, but the theory of colonisation on which South Australia was founded did not provide for this contingency. The politicians would eventually have to decide whether to keep putting up land for sale until the farmers failed, or to assess where the survey should stop. Once it was acknowledged that the northern country would never support agriculture, the drought-adapted pasture of 'the beautiful salt bush plains', as they were later remembered, would be spared the plough.[20] The pastoral leaseholders could consider their occupation permanent and begin to make long-term investment in their runs. If it could be determined how far the agricultural land extended, in a way that was clearly objective and impartial, competition between farmers and pastoralists could be forestalled. Goyder wanted to find a solution for the sake of the colony and the well-being of those directly affected, but he also knew that he would have to deal with the matter in any case. The colony's founding precepts ensured that the Surveyor-General exercised a great deal of influence on the course of settlement through the land use decisions that inevitably had to be made in a system where survey preceded settlement. The attainment of self government in 1857 had reduced the position of Surveyor-General from a member of a small council advising the governor to head of a department under an elected minister (or commissioner in this case), but early governments were short-lived. The succession of commissioners could not help but rely on advice, and the post of Surveyor-General remained a position of considerable influence.

Early in 1865, as the north was beginning to wither and parch, a parliamentary committee was formed to investigate the sale of crown land. Discussing the extent of the remaining agricultural land, it was obvious that Goyder had a clear vision in his own mind of how far that land extended. He assured the committee that he would have 'no difficulty at all' in preparing a map that would distinguish the pastoral and agricultural country, and supported the idea of dividing them.[21] When asked if he could fix the boundaries of the agricultural lands, Goyder replied: 'I think I can fix the boundaries by what would be the reliable rainfall'. He then defined this boundary as: 'the apex of a cone; Mount Browne [*sic*] forms the apex, and the line comes down south, and very nearly south-east — spreading out to the eastward, then dropping to the south again'.[22] Particular points of reference were identified. Goyder also spoke of a transitional realm — now the 'marginal country' — between the 'extreme' runs, as he put it, and safe agricultural country, and tied the line explicitly to the agricultural practices of the time. This forerunner was simpler in shape, and reached further north at its single peak — Mount Brown lies beside the head of Spencer Gulf — than the line he would actually produce (see map[23]).

The idea of such a line was not novel in itself. As early as 1858, the prominent pastoralist and parliamentarian, George Hawker, described a line from Mount Bryan to Mount Remarkable (the eastern shoulder and western tip of Goyder's Line) and down to Port Lincoln, on the Eyre Peninsula, beyond which grazing was no longer a matter of shaking nuggets from the golden fleece.[24] In 1865, six days before Goyder offered to map the reliable rainfall, a meeting of thirty representing the northern squatters resolved to ask the government for a line defining the drought-affected country, and for an extension of tenure beyond it.[25] It was this request that ultimately sent him back into the waterless north in November that year, exhausted as he was by continued work on the valuations. His instructions were simple: to 'lay down on the map, as nearly as practicable, the line of demarcation between the portion of the country where the rainfall has extended, and that where the drought prevails'.[26] But as he later explained:

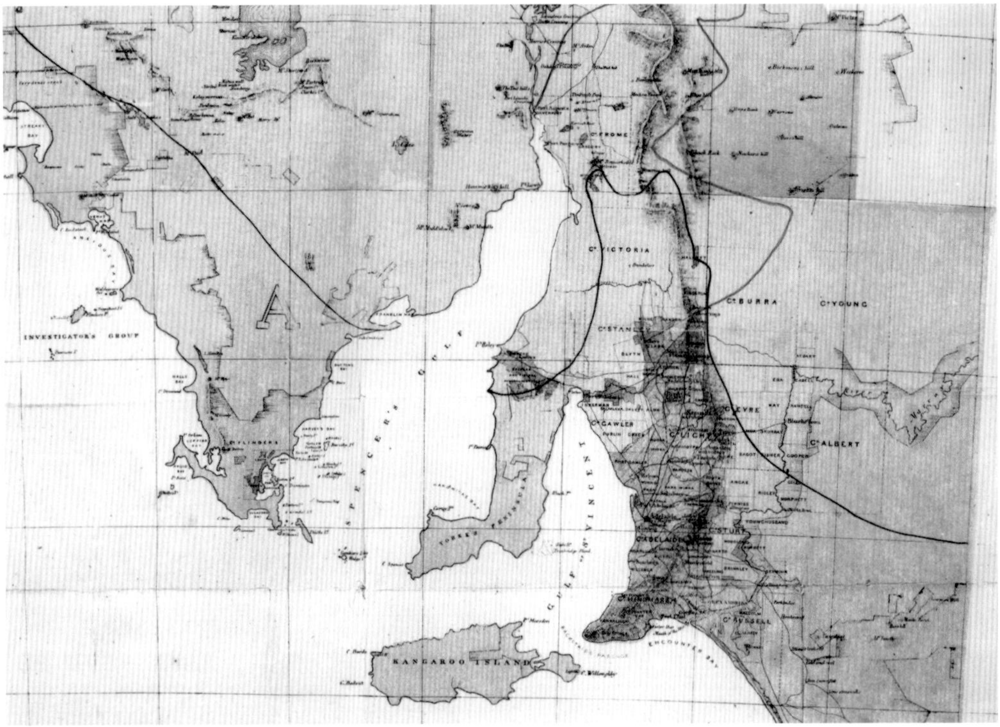

Goyder's Line as presented in February 1866, in a section from Map of the Northern Runs.

LA TROBE COLLECTION, STATE LIBRARY OF VICTORIA

> Had the drought ... been that of an ordinary nature, there had been no necessity for my leaving town upon this duty, as the line of demarcation might have been shown from information previously in my possession ... The drought, however, being of an unusually severe nature, and extending more generally than was previously known, it became indispensable to add to my previous experience the knowledge of the state of the country as it now exists.[27]

Unfortunately, the pages covering this journey were torn from the field notebook — a unique case of such damage — so no record remains of Goyder's observations, apart from a note about the presence of dead tall trees as indicators of drought at long intervals.[28] In a memo defending a second map, on which further lines defined different schedules of drought relief, he explained that the lines were the result of 'personal experience and knowledge of the country, obtained by frequent visits extending over a series of years and a variety of seasons'. The experience expressed in the lines was of 'reliable rainfall'.[29] Goyder had not simply mapped a drought border, as directed, but had used the situation of the drought to continue to pursue the longer term objective of accurately dividing the pastoral from the agricultural land by reference to rainfall variability, though no one appears to have noticed, or cared.

Even without the field notebook, it is possible to look at the line as a whole and see how it was constructed, because although it expressed a single principle, it was evidently not created by the application of a formula. Goyder's experience of the country behind Fowlers Bay would have determined the western end of the line, and at the point where the line meets the coast of the Eyre Peninsula, a large area of hilly country was leased as pastoral land, making it possible that information had been obtained from the lessees. The line runs just below it. The eastern curve connected the Murray River at Swan Reach with 35° S on the Victorian border excluding the waterless mallee country, known as the 'monster scrub', in the process. (In 1865, techniques for large-scale clearing of mallee had not yet been developed, and mallee country, or 'scrub', was unsuitable both for agriculture and pastoralism. This section of the line was later adjusted.)

The country really at issue surrounded the central section of the line, which can also be considered section by section. The western section, by the coast, excludes what was largely mallee. The two 'horns' enclose Mount Remarkable on the west, and the hills of the South Flinders Ranges to the east. Between them the line dips down to exclude the Willowie Plain. Before this was cleared in the 1870s, the south-eastern section contained large areas of mallee with patches of native pines. There was bluebush in the north-east and saltbush in the west.[30] The eastern shoulder of the central section takes the line out to just beyond the edge of the eastern-most ridge of hills. A squatter named Alexander McCulloch, a serious sheep farmer who made a close study of the local rainfall, is reputed to have helped Goyder fix the line in this area.[31] (The only people to have taken Goyder's point seem to have been squatters with holdings in the outside country, but any connection with pastoralism damned the line in the eyes of farmers. The valuation for McCulloch's run, Gottlieb's Wells (near Terowie), mentioned rainfall.) Below this shoulder, the line now takes the form of a north–south road beside the hills, a real and evident border between the wheat country and the saltbush. This part of the line has never been disputed, because the country to the east lacks both the rainfall and the soil for agriculture.

Dividing pastoral and agricultural country

The line's real purpose was made public in 1870, when Goyder reported on Victoria's experiments with free selection. Reflecting on the competition for land between the farmers and the squatters, and on the responsibilities of government, Goyder promoted the idea of a division. In this respect, South Australia was apparently fortunate. There, he explained, 'nature has clearly established a line of demarcation beyond which permanency of pasture may be given to the pastoral tenant, without detriment to the agriculturalist; and a certain portion of Victoria, viz, the north-west, partakes of the same character ... '[32] Land legislation was a matter of compelling public interest at the time, and the report was reprinted by the *Advertiser* and sold as a pamphlet for a penny.

Unfortunately, while the report presented Goyder's understanding of the purpose of the line to an interested public, it did not contain any account of the line's real meaning. At about this time, the English novelist Anthony Trollope was on a tour of the Australian colonies, and the line had the cultural distinction of being described by him, in his book *Australia*.[33] The accuracy of Trollope's account, which begins by referring to the frequency of drought in the inland and properly outlines Goyder's concern both for agriculture and pastoralism, suggests a meeting between the two.

When the line was dismissed as arbitrary nonsense four years later, the farmers were advancing into country in which tens of thousands of sheep were known to have died of hunger and thirst in the previous decade, and no mere warning was likely to have stopped them. Along with the collective frenzy characteristic of booms and rushes of all kinds, the advance into the north was driven by specific local factors — the inexperience of new settlers, the false confidence of second generation farmers from the south, and the destructive attitude of 'surface miners', who cropped the soil until it was exhausted and moved on. There had also been a history of proclaimed limits, from the beginnings of settlement, which had been proved false. When political expediency led to a profound change in style of government, from paternalistic protection to one of free experimentation at the farmer's risk, not all farmers understood this, and some continued to believe that land made available for agriculture had been judged suitable.

At the same time there were two theories abroad which discounted the lessons of experience. One was the folk belief referred to by Goyder, that 'rain followed the plough'. The second was a theory developed by experts and championed by the environmentally concerned. The connection between trees and the atmosphere was coming to be understood, and around the world it had been observed that clearing trees created deserts. It was concluded that if cutting trees produced desert, planting them would banish it — that human beings could (and naturally should) control and change climate, by planting trees to produce rain. Against theories like these, and the careless destructiveness of transient farmers, even a properly understood line of reliable rainfall would not have prevailed.

Nineteenth-century communications and Goyder's position as a public servant no doubt hindered the dissemination of his message, but the real reason for the still birth of the line of reliable rainfall, and the lusty vigour of its twin, Goyder's Line, seems to have been the simple inability to think in terms of rainfall variability as an aspect of climate. This was concealed by the misleading familiarity of the words 'unreliable', 'uncertain' and 'precarious'. People unthinkingly assumed they had fully comprehended what Goyder had to say, though all that seems to have been retained was a sense that rainfall in the north wasn't good enough. Before the Land Commission of 1888, Goyder explained his line in terms of drought, and 'reliable' or 'precarious' rainfall, and testified again to the dramatic changes in the vegetation of the north. He was then asked if he had pointed out 'the deficiencies of the rainfall'.[34] This failure to mirror his language was typical, and not insignificant.

A demonstration of this cultural incapacity at work is provided by John Ednie Brown, a forester from Scotland. Brown had been recruited to the newly created post of conservator of forests by Goyder, who was chairman of the Forests Board, but the two were soon in conflict over the climate of the north. Brown believed that planting trees north of Quorn would bring rain to the inland. Goyder argued that trees could not modify climate on this scale, and cited the presence of vast areas of mallee, which he recognised as forest adapted to aridity and drought, as evidence. It was lack of rain that stopped the trees from growing, he argued, not lack of trees that stopped the rain from falling, but he was criticised for asserting himself where he lacked expertise.[35] Brown, who had made only one cursory tour of the north, wanted to go ahead with his plan to improve the inland climate, but because he was dealing with Goyder directly, he could not escape coming to terms with the uncertain rainfall in some way. In his report for 1881–1882, Brown described a method of tree planting designed to circumvent initial difficulties. Rather than plant seedlings, seed was to be sown into ploughed ground, there to wait safely until rain came to trigger germination and 'nourish the young plants in nature's own time'.[36] Brown's imagination could go only so far before it defaulted to the vision of a benign, attentive, maternal 'nature', providing rain when it was needed.

At about the same time, and as his warnings were being vindicated, Goyder protested at his line being 'commonly, though erroneously, called Goyder's line of rainfall', but without explaining why, and without any effect.[37] After his death in 1898, the line of reliable rainfall was entirely lost, though it came close to being resurrected in 1918 by Griffith Taylor, the geographer who was derided for his unpatriotic acknowledgment that Australia was a continent beset by drought. In his book, *The Australian Environment (Especially as Controlled by Rainfall)*, Taylor declared that the line 'practically agrees with the southern boundary of the distinctly characteristic "salt-bush, mulga, and dwarf mallee country"', and noted that although an immense amount of wheat could be grown beyond the line in favourable years, the 'safe farming country lies south of this ecological isopleth'. The term 'safe' referred to rainfall reliability, so Taylor had all the elements in hand, but he went on immediately to explain that when the isohyets were adequately charted, the line was found to agree with the 12-inch isohyet. He recommended this 'logical method of estimating the possibilities of a new country' for the arid lands.[38] By the early 1960s, the line's meaning had become so obscured that when Meinig refused to agree that it resembled any isohyet of average rainfall, he was forced to describe it instead as a 'very personal' geographic concept, the local importance of which was borne out by its cartographical persistence.[39]

From then on the situation gradually improved, but although references to reliability appeared, unheralded, the basis of Goyder's celebrated foresight remained essentially mysterious. Now, anxiety focuses our attention on El Niño, La Niña and the Southern Oscillation Index, and a 'Climate Variability in Agriculture Program' has been called into existence. There is widespread awareness that many of Australia's plants and animals are adapted to variability as well as aridity. The gap across which Goyder mouthed his warnings has finally closed and the line of reliable rainfall can emerge at last, if only into history. If, as one obituary writer supposed, Goyder's name would never be forgotten because of his famous line, then perhaps it will survive as a cultural mnemonic for a dominating and dramatic aspect of Australia's — not just South Australia's — climate, and for the value of attentiveness to the natural environment, and the willingness to learn.[40]

Janis Sheldrick

Janis Sheldrick is an independent scholar who has been researching and preparing a biography of George Goyder since 1993. She received a PhD from Deakin University in 1999 for parts of the work covering Goyder's early journeys and the origins of the line, and the Northern Territory survey.

6

Rainmaking in CSIRO
the science and politics of climate modification

R. W. Home

Drought, or the threat of drought, has been a constant accompaniment of Australian life, a recurring experience for the European invaders as it has always been for Aboriginal Australians. Settlers enticed to take up land in the interior by its lush appearance in a good season regularly found themselves facing disaster as they discovered that good seasons were all too rare in most parts of the country. It took years of experience before weather patterns were recognised, and by then it was often too late — people had already committed whatever resources they could muster to making a go of things. The image of the desperate Aussie battler and his long-suffering wife confronting the endless summer heat on their parched selection, helplessly watching their stock die as they wait for the rain that never comes, is a familiar trope. So, too, is the image of their fighting against adversity, of overcoming the worst that Nature can do to them by sheer hard work that reshapes the potential of their land. Still today one sees, after every drought, more and more dams being dug and more and more bores being sunk in an effort to secure more dependable supplies of water for when the next drought comes.

In the aftermath of the Second World War, Australians looked to the future with new optimism and a newly developed faith, deriving from what had happened during the war, in the ability of science and science-based technology to solve whatever problems confronted the nation. Until the war, Australia had relied on overseas suppliers for most of the manufactured goods it needed. Under the pressure of war, with the help of Australia's scientists and engineers, ways had been found to manufacture many of these goods locally, and now the attractions of a new and more self-sufficient, industrially based economy beckoned. Australians also looked to science to transform the land. Already, before the war, the scourge of prickly pear had been conquered, while the triumphs of the wheat breeders in producing new varieties suited to Australian conditions were well known. Now, vast areas of previously marginal country were being rendered productive by the addition to the soil of traces of various elements, the importance of which had been identified by Australian scientists in the 1930s.

Large-scale research was promised to underpin the development of the continent's 'empty north', and also to protect the wool industry, so vital to the nation's economy, from the looming threat posed by the development of artificial fibres such as nylon. Plans were being developed to divert the waters of the Snowy River into the inland river system of the Murray-Darling Basin. Most exciting of all, even if also the most worrying if the secret fell into the wrong hands, the energy of the atom that had been released with such devastating effect over Hiroshima and Nagasaki promised to usher in a new and expansive era when brought under control and applied to peaceful purposes. For a nation that was thought at the time to be bereft of oil and to have only very limited supplies of coking coal, a golden future based on effectively unlimited supplies of atomic energy beckoned. In addition, atomic power might transform the landscape: in the years following the war, the possibility was seriously bruited of using atomic explosions to blast out new harbours on Australia's north-west coast or, even more ambitiously, to open a channel from the head of Spencer Gulf to Lake Eyre and flood a great basin in the centre of the continent that is, conveniently, already below sea level, thereby producing vast changes in climate patterns across the whole country.

It is within the context of such grand ambitions, linked to the application of science to national development, that we need to see the new research programs in atmospheric physics on which the nation's chief scientific agency, the Council for Scientific and Industrial Research (CSIR, later the CSIRO) embarked in the immediate post-war period. The new programs likewise reflected a strong sense among CSIR's senior management that the existing Commonwealth government agency concerned with the weather, the Commonwealth Bureau of Meteorology, was too restricted in its operations to collecting data, monitoring weather patterns and issuing forecasts, and too little committed to fundamental research into the physics that lay behind the patterns that were observed. These concerns led to negotiations with the Bureau culminating in an agreement that the Bureau was not in a position to undertake fundamental research into the physics of the atmosphere, and that CSIR should do so instead.[1] Within CSIR, there was a characteristic confidence that benefits would eventually flow from such research: 'The importance of weather to Australia undoubtedly justifies a much more intensive study of the physics of the atmosphere', the Council declared, 'as only a knowledge of the basic physical phenomena can ultimately lead to the possibility of making successful long-range weather forecasts'.[2] Soon, much more grandiose ambitions were being entertained that ran well beyond better forecasting to actually modifying Australia's weather by increasing the rainfall in drier parts of the continent.

The principal driving force behind CSIR's push into atmospheric physics was F. W. G. White (universally known as 'Fred'), a New Zealand radio physicist who had served as chief of CSIR's secret Radiophysics Laboratory during the latter years of the war before being appointed in 1945 as a full-time member of the CSIR Executive. Under White's leadership, CSIR embarked on a two-pronged drive to develop a better understanding of the physical processes underlying the nation's weather. On the one hand, a talented young English meteorological physicist, C. H. B. Priestley, was hired to head a new section (later a full-scale division) of

Meteorological Research that built up a long-term program of micrometeorological investigations. On the other, the Radiophysics Laboratory, now no longer secret but transmogrified under its new chief, E. G. Bowen, into the Division of Radiophysics, would undertake research into the physics of clouds. It is with the latter project that I am concerned in this paper.

The Radiophysics Laboratory

The Radiophysics Laboratory had been established in 1939 in the grounds of the University of Sydney, under conditions of great secrecy, in order to build up a cadre of Australian expertise in a newly developed technique of great military significance known as radio location, or radar. The founding chief of the laboratory was Australia's leading radio physicist, David Martyn, but in 1941 Fred White had been brought in over his head to take control — a development that Martyn bitterly resented. During the war, the laboratory grew to become one of the nation's largest research groups, with an outstanding record of research and development of radar-related equipment for the armed services. As the end of the war came in sight, the decision was taken to maintain the laboratory as a government-funded centre for Australian research on civilian applications of radio and radar.

When it came to selecting peacetime projects to pursue, the laboratory benefited greatly from the new air of freedom that was sweeping over CSIR. Prior to the war, CSIR's leaders — most notably the long-serving chief executive officer, David Rivett — had chafed at the extent to which the research being done had had to be confined to solving practical problems. Then of course during the war there was no question but that all efforts had to be focused on urgent, war-related problem solving. Yet Rivett and his colleagues were convinced that, in the long run, such an exclusive focus on short-term issues would stultify creative research, and they had long looked forward to a period when CSIR's scientists would be able to devote a significant proportion of their time to fundamental investigations. The coming of peace, the new standing that science enjoyed within the Australian community as the likely key to securing a better future, and access to new funding as a result of the dramatic wartime increase in the financial power of the Commonwealth government combined to give them their chance. Henceforth, it was declared, instead of focusing solely on solving practical problems, CSIR scientists would be expected to devote a significant part of their time to fundamental research.[3]

Planning for the Radiophysics Laboratory's transition to peacetime work began long before the war ended, and continued into the peace. Meteorologically orientated work consistently featured in reports proposing programs of future work. E. G. ('Taffy') Bowen was responsible for the two most important reports, prepared respectively in July 1945, when he was deputy chief of the division, and August 1946, a year after the war ended, by which time he had become chief and the program he had outlined a year earlier had been largely brought into being.[4]

In the first of these reports,[5] Bowen set out three 'general principles determining the choice of programme' that clearly enunciated the new spirit that was sweeping CSIR. These were 'the active pursuit of fundamental problems in radio and radar', 'the selection of applied problems, the solution of which will be of greatest utility to the people of Australia and make best peacetime use of the techniques in which the

"Congratulations, sir, on a most successful experiment."

MOLNAR FAMILY/NATIONAL LIBRARY OF AUSTRALIA

laboratory has become competent', and 'the maintenance of a healthy balance between fundamental and applied work'. The laboratory's specialised techniques, it was noted, could 'make a practical contribution in many fields of fundamental investigation'.

Several lines of investigation that had been pursued in the laboratory during the war converged on questions of atmospheric physics; for example, the phenomenon of 'super-refraction' — the discovery, obviously of immense practical importance, that microwave radar signals could be propagated far beyond the horizon thanks to atmospheric refraction — had been shown to depend, in ways that were not yet fully understood, on temperature and humidity gradients in the lower atmosphere. Again, microwave radar sets had proved capable of detecting storms at distances of up to 150 miles — due, it was thought, to the scattering of radio waves by water drops and ice particles — while signals had also been found to be reflected, in ways that were not at all well understood, from discontinuities in the middle atmosphere. Research aimed at developing more reliable radar systems was thus going to be inextricably linked to meteorological issues; conversely, Bowen pointed out, 'the study of the physics of the atmosphere can benefit considerably from the use of radar techniques'.[6]

The document that Bowen drew up a year later was, in effect, a report on the progress that had been made in the intervening period. 'It will be observed', Bowen now remarked, 'that almost half of the work of the laboratory is of a fundamental character, rather less than half comprises the application of radar techniques to the peacetime needs of Australia, and only a small proportion is devoted to work of direct interest to the Armed Services'.[7] This represented a remarkable transformation for the laboratory, because during the war its program had been set entirely by the needs of the Services. So far as the peacetime application of radar techniques was concerned, the emphasis was on developing radar equipment for aircraft navigation and air traffic control, and radar aids to surveying. The civil aviation equipment, in particular, involved mostly straightforward adaptations of equipment originally developed for military purposes; rapid progress was being made, and already Bowen was foreseeing its coming to an end.[8] In addition, work was being done in conjunction with the National Standards Laboratory to develop electrical measuring techniques and physical standards at radio frequencies. In the longer run, however, such work would become the responsibility of the National Standards Laboratory alone. Much effort had also been put into developing the facilities and skills needed for undertaking research in thermionic and vacuum physics, above all in connection with building a linear accelerator for atomic energy research.

The major focus of the laboratory's work continued, however, to be what in the report was termed 'Radio propagation and atmospheric physics research'. This included further studies on super-refraction and the temperature and humidity structure of the lower atmosphere, and on the scattering of radar signals by rain and clouds. Since clouds were relatively transparent to radar signals, the techniques being developed permitted study of processes going on within them. 'Radar', Bowen proclaimed, 'will provide an elegant tool for studies in atmospheric physics and the problems associated with the formation of rain'.

The section of Bowen's progress report dealing with propagation also gave news of some dramatic discoveries of a different kind, the pursuit of which was soon to transform the overall balance of the laboratory's work. A year earlier, in a section of his report headed 'Radio Noise', Bowen had noted that 'receivers are now sufficiently perfect to cause the ultimate limit of receivable signal to be set by the level of the noise picked up on the aerial'. Finding out more about the sources of such noise was therefore of practical importance. Among these, Bowen noted, in a limited range of the radio spectrum, 'a type of noise appears which is thought to originate in the stars or in interstellar space'. While little was known about this at that time, things moved very quickly thereafter. In December 1945, Fred White reported to the CSIR Executive after a visit to Sydney that, while trying to detect radio frequency radiation from the sun, a group in the laboratory had obtained 'physical results of very great interest'.[9] A letter had been written to *Nature* reporting these results and a 'fairly comprehensive paper' was being prepared.[10] As Bowen put it:

> The subject is clearly of fundamental importance to astrophysics since it provides a method of investigating extra-terrestrial phenomena by means other than light. The work is still in its infancy but the success already achieved gives strong grounds for hoping that important astrophysical discoveries relating to the structure of the outer atmosphere of the sun and stars will derive from further study.

In fact, the study of 'cosmic noise' — radiation at radio frequencies emanating from celestial sources — quickly became a major strand of the laboratory's overall research program, with the group responsible for the work making a series of famous discoveries that catapulted them, almost overnight, to world leadership in the emerging science of radio astronomy.[11] This was exciting stuff. Leading the world was a heady experience for CSIR, locked as it had traditionally been into problem solving for Australia's rural industries. It was an experience to be savoured. Yet the cosmic noise group's very success created dilemmas of its own. To be sure, astronomy was a science that had traditionally attracted public interest and support, but it was nevertheless not an obvious focus of attention for a government-based research organisation set up to assist the nation's primary and secondary industries. Moreover, the discoveries the group was making took their work further and further into the realm of 'pure' research where it became ever more implausible to argue the prospect of practical spin-offs, even in the long run.[12] Given the successes that were being achieved, however, it would be inconceivable to cut the research short or give it away. Somehow or other, it would have to be accommodated within the Radiophysics Division's overall research program.

Meanwhile, the atmospheric physics work was also making rapid advances. In this case, the spur came from America, where in November 1946 Vincent Schaefer at General Electric, following up on some laboratory observations, had 'seeded' a cloud with dry ice and then watched snow fall from its base. Already by early December E. B. Kraus and Patrick Squires at Radiophysics had successfully repeated Schaefer's initial laboratory procedure and drawn up a scheme for a suite of investigations on 'artificial stimulation of rain'.[13] Their first field experiment, in February 1947, was strikingly successful: after dry ice was dropped on a large cumulus cloud, the cloud changed form dramatically and heavy rain began falling from it. The possibility of artificially stimulating the production of rain had been demonstrated, and gave rise to the dream of modifying Australia's notoriously harsh, dry climate. If this were to be done in a controlled way, however, a much better understanding would be required of the actual process of rain formation in clouds.

And so a new research agenda was established for the cloud physics group at Radiophysics, one that was to dominate the group's work for many years. This combined practical experiments on cloud seeding, to establish the most efficacious method of doing this, with fundamental research into the internal processes within clouds that led to the formation of rain. The prospect was, if anything, even more exciting than the work on cosmic noise. As Bowen wrote for the section devoted to the Radiophysics Laboratory in CSIR's annual report for that year, 'Initially, these subjects [that is, "radio meteorology"] were studied mainly from the radio point of view, but the meteorological problems have become so interesting that they now form the principal part of the study'.[14]

By the time the next CSIR annual report was compiled, a year later, rainmaking had clearly become the paramount objective, and more fundamental research on cloud physics was no longer being justified for its own sake but rather for the underpinning it provided for rainmaking.

During the course of the following few years, the range of research activities being pursued in the Radiophysics Laboratory narrowed considerably. Some projects were successfully completed, others lapsed or — as in the case of the early generation electronic computer, CSIRAC, that was constructed in the laboratory and commenced operations in 1951 — were passed over to other organisations. The net result was that from the mid-1950s virtually all the projects that had once been identified as constituting the 'applied' side of the research program in Radiophysics had been more or less abandoned. More and more, radio astronomy and cloud physics became the only fields of research being pursued. Both continued to make rapid advances, and both continued to enjoy very strong support from the CSIRO Executive, especially from Fred White who became the organisation's chief executive officer when it was restructured in 1949 and later its long-serving chairman. The radio astronomy group, in particular, was one of the top two or three groups in the world in what had become a fast-moving and glamorous new field of scientific inquiry. Given, however, that the radio astronomers' work continued to be viewed as 'fundamental', the Division's overall research profile could only meet CSIRO's general guidelines for balancing fundamental and applied research if the cloud physics work was redefined as 'applied', although it had initially been classified unambiguously as fundamental research. This it was now plausible to do — but only so long as the rainmaking side of the work, rather than the long-term investigations of cloud dynamics that had initially been envisaged, was seen as paramount. In other words, in addition to the sheer excitement engendered by the prospect of making rain when it was needed, there were clear institutional and political reasons for rainmaking to be the principal focus of the cloud physics group's research.

The point is worth emphasising, because — responding, no doubt, to the eventual collapse of the rainmaking program — histories of the cloud physics project have tended to lay heavy emphasis on the group's achievements in fundamental research, and to portray the rainmaking work as a mere appendage to this. Indeed, according to the most recent account, the scientists quickly became disillusioned with the rainmaking work and persisted with it only because the powers that be insisted they do so.[15] To be sure, the growing number of calls for assistance from drought-affected communities around Australia was a distraction for the group since they saw their role as doing the underlying research that was necessary, not becoming service providers. Moreover, they were not yet confident about their techniques and in any case did not have the resources to respond adequately to appeals for help. But the fact is that Bowen, enthusiastically supported by White, even though he was fully aware of the political necessities that constrained research in CSIRO, deliberately wound down other areas of potential 'applied' research and left the Division with just its two research programs in radio astronomy and cloud physics. His doing this testifies to the confidence he and White had in the viability of the 'applied' side of the cloud physics research — that is, the rainmaking — and to their primary commitment to this aspect of the work.[16]

Cloud physics and rainmaking

The early years of the cloud physics project saw a rapid build-up of staff and supporting infrastructure (including a specially fitted out DC3 aircraft supplied by the Royal Australian Air Force). Ingenious instruments were developed for measuring various aspects of the internal dynamics of clouds including, crucially, the number and sizes of the water drops that were present. Radar was an important tool, both for measuring ice crystal or drop size and for observing the changes within a cloud after

"Faster! The damn thing will rain before we make it."

MOLNAR FAMILY/NATIONAL LIBRARY OF AUSTRALIA

it had been seeded. The work focused particularly on the mechanisms involved in the formation of natural rain — that is, the formation and growth of droplets in clouds — to underpin attempts at stimulating rain artificially. Until this time, it had been widely accepted that ice crystals were generally required to serve as condensation nuclei on which the drops that were big enough to constitute rain formed, and considerable effort was devoted to studying how this happened. Dropping dry ice into clouds was thought to stimulate the process. Now, however, it was observed that rain could fall from clouds even when they were entirely above the freezing point and so contained no ice crystals. The Radiophysics group contributed significantly to the study of droplet growth by coalescence in non-freezing clouds and, more generally, to a better understanding of droplet formation and of the internal dynamics of clouds. They quickly won international recognition as one of the leading groups in this challenging new field of research, and with some justification were soon able to claim that 'while related work has also been proceeding in other countries of the world, it is largely as a result of investigations carried out in Australia that the basic problems involved are now more clearly understood'.[17]

Meanwhile, research was also proceeding on problems more directly related to rainmaking. Initially, further seeding experiments with dry ice were carried out, using a downward-looking radar set in an aeroplane hovering over the seeded cloud to track the formation and descent of the snow, ice and raindrops that were produced by the seeding.[18] The process quickly came to be well understood. Following the recognition that rain could also be produced from non-freezing clouds, attempts were made to stimulate rain by spraying water into a suitable cloud.[19] Water droplets released at the base of a growing cloud would be carried upwards by the air currents, growing as they went through collisions with the small water droplets of the cloud until, having grown so large that the updraft could no longer sustain them, they fell as rain. However, both of these processes required inconveniently large quantities of seeding material — either dry ice or water, as the case may be. A search for alternative seeding materials culminated in the American discovery that comparatively small quantities of silver iodide could act as a highly effective seeding agent in freezing clouds. Rainmaking research in Australia, as elsewhere, concentrated thereafter on the use of silver iodide.

The standard technique that emerged involved burning the silver iodide to produce a fine smoke of silver iodide particles that was then injected into a freezing cloud. In America, where numerous commercial rainmaking ventures sprang up in the wake of the experiments of Schaefer and his colleague, the Nobel prize winning physicist Irving Langmuir, spectacular claims were made about the effectiveness of smoke generators set up on the ground, the smoke from which was carried up into the cloud by ordinary convection processes. The publicity that was generated served to raise public expectations in Australia, as elsewhere. The Radiophysics group undertook extensive research on the use of American-style ground-based smoke generators before concluding that there was no evidence that smoke from them ever reached the clouds, let alone caused these to drop rain. Furthermore, in daylight, photochemical reactions were found to reduce rapidly the effectiveness of the silver iodide as a seeding agent, so that it was necessary for the smoke to reach the cloud promptly if it were to work. In the American experiments, even if some of the smoke had reached the cloud, it was unlikely by then still to have been effective. The Australian group therefore concluded that if the seeding were to work, it would have to be done from an aircraft (or perhaps from a mountain top): a smoke burner they devised that could be slung under the wings of an aeroplane, that burned a solution of silver iodide and ammonium iodide dissolved in acetone, worked extremely well and became internationally standard equipment.

It was also important to determine what kinds of clouds were suitable for seeding. It was found that, for reasons not at all well understood, there were important differences between clouds formed over land and over sea, with many more but much smaller droplets in continental clouds. Since larger droplets are needed if rain is to form, this meant that continental clouds were much less apt to release rain than were maritime ones, a discovery of considerable significance to would-be Australian rainmakers. Another crucial factor was found to be the temperature of the upper levels of the cloud; with both cumulus and stratiform clouds, provided this temperature was less than –7°C, seeding would inevitably be followed by precipitation within 20–25 minutes.[20] On the other hand, if the temperature were less than about –15°C, rain would almost certainly fall, with or without seeding. To judge the viability of a seeding program, it was important to establish that seeding made a difference — that is, that it resulted in rain from clouds that would not otherwise have yielded it. Randomised trials in which single clouds were seeded while neighbouring ones were not gave encouraging results. An extensive series of area trials was therefore begun in 1955 and continued for many years, to find out whether the rainfall in a specified area could be increased by seeding, by comparison to a nearby control area where the clouds were not seeded. At the same time, those involved found themselves frequently being called upon to help alleviate drought in one part of the country or another, and in response to this, various other *ad hoc* seedings were undertaken.

The results obtained can only be described as mixed. Seeding over the Snowy Mountains seemed to produce a significant increase in rainfall, whereas in other cases — in South Australia, and in the New England region and in the catchment area for the Warragamba Dam in New South Wales — there seemed to be no such increase. More puzzling — even alarming — was an apparent decrease in rainfall, in all areas tested, over time. Furthermore, even the apparently favourable results from the Snowy Mountains were tarnished by a dispute that broke out over how the measurements were to be interpreted, with the CSIRO group initially claiming a 20 per cent increase in rainfall in the seeded area and the Snowy Mountains Hydroelectric Authority, while acknowledging that there had probably been an increase, taking a much more cautious view on how large this might have been.

In experiments of this kind, statistics are all-important — very difficult statistics, given that we are dealing with such an inherently variable thing as weather — and experiments have to be extremely carefully designed if the conclusions reached are to carry conviction. Unfortunately, many of the early Australian experiments, like those done elsewhere, were designed more to demonstrate possibilities than anything else, and so did not include adequate statistical controls. Some outside observers, sensitised no doubt by the exaggerated reports of rainmaking successes coming from America, were inclined to adopt a highly sceptical view from the beginning towards the claims being made. But I think it also has to be said that Bowen brought some of the trouble on his own head by adopting what to his critics seemed a cavalier attitude to statistics. This was shown most clearly, perhaps, by his 1953 suggestion that there was a previously undetected cyclical pattern in world-wide rainfall figures that he claimed was linked to a cyclical variation in the amount of meteoritic dust in the atmosphere capable of acting as raindrop condensation nuclei. This depended in turn, he supposed, on the Earth's regularly passing through various meteor showers in its orbit around the sun; the rainfall peaks he had detected occurred, he said, twenty-nine or thirty days after the Earth passed through a meteor shower, this being the time required, he supposed, for the dust to find its way down to the troposphere.[21] The proposal attracted a storm of criticism that focused especially on the statistics on which Bowen had depended. Notable attacks came from elsewhere within CSIRO — from W. C. Swinbank at CSIRO's Section of Meteorological Physics, which throughout these years carefully kept its distance from the rainmaking work, and from Bowen's old enemy, David Martyn, now based at Mount Stromlo Observatory.[22]

Others less personally involved were, however, equally critical; for example, the director of the Jodrell Bank radio observatory in England, A. C. B. Lovell, a leading expert on meteor showers, told Bowen he found his hypothesis 'extremely improbable'.[23] Bowen, however, was unrepentant and continued to promote his meteoritic dust hypothesis for several years while seeking new evidence in its favour.

While this controversy was not directly linked with the rainmaking work of Bowen's Division, it surely had an undermining effect on the general credibility of the rainmaking program with which Bowen was so closely identified. The same can be said for the criticism later levelled by David Martyn at Bowen's suggestion that a lunar effect he claimed to have found in rainfall patterns might be explained on the basis of the Moon modulating the amount of meteoritic dust reaching the Earth.[24] The controversy with the Snowy Mountains Authority was, however, much more damaging since it related directly to the credibility of the rainmakers. So, too, did a later controversy over the interpretation of data from a seeding experiment in the Wimmera-Mallee region of Victoria in 1966.[25] By this time, although those most closely associated with the rainmaking program continued to depict it as a normal, successful scientific project, a growing number of critics were viewing it more dubiously. Among these was the Commonwealth Director of Meteorology, W. J. Gibbs, who in 1965 issued a public statement on cloud seeding that by inference damned CSIRO's efforts. It was, he said:

> difficult to determine whether fluctuations in rainfall that occur at the time of cloud-seeding were produced by seeding or would have occurred naturally ... There is some doubt whether a statistically significant answer is attainable but it is obvious that if such an answer is to be found it will only result from a carefully designed, operated and evaluated scientific experiment. ... Cloud seeding in Australia has been of three types — 'single cloud experiments', 'controlled experiments' aimed at assessing the effect of seeding on the rainfall over an area, and 'ad hoc operations'. The results of 'single cloud experiments' suggest that cloud characteristics can be modified by seeding. CSIRO believes that rainfall has been increased in some of the 'single cloud experiments' but the evidence available to the Bureau of Meteorology is not sufficient to form an opinion on these results. The 'controlled experiments' over designated areas have not yielded a conclusive result. The 'ad hoc operations' cannot possibly do so.[26]

White retired in 1970, followed by Bowen a year later. Bowen's successor as chief of Radiophysics, the eminent radio astronomer Paul Wild, quickly had the whole cloud physics project moved out of the Division. Ten years later, the rainmaking side of the work was officially abandoned, for both scientific and economic reasons. 'The scientific evidence', it was said at this time, 'is made up of two factors — firstly, the increasing realization that cloud conditions suitable for seeding are far less frequent than previously thought, and secondly, the inherent natural variability of rainfall in Australia'. The latter phrase indicates an acceptance, at last, at least in the scientific community, that Australians had to learn to live with drought and adjust their practices accordingly, that science was not going to provide a 'magic bullet' to change the continent's climate. There was, however, a further consideration:

> for cloud seeding to be worthwhile in rural areas, it must cost less than the economic benefits, in terms of additional rural production, which result from any increases in rainfall due to the seeding. Unfortunately, the trend is the very reverse. The cost of flying and maintaining aircraft to carry out cloud seeding is increasing much faster than the returns on rural products.[27]

Yet the fact that it took ten years to reach this decision, even after White and Bowen relinquished control, speaks volumes for the support the project enjoyed both within CSIRO and in the wider Australian community. To be sure, this support had been carefully fostered over the years, but the project also clearly resonated with attitudes that were widely held. Neither White nor Bowen was originally Australian, but they had tapped into a persisting Australian preoccupation with reshaping the environment. By the same token, neither man was known for making decisions in relation to science on sentimental grounds: on the contrary, they shut down other research projects without hesitation if they judged them not to be going anywhere. It was not inevitable that cloud physics should become a major strand of the research program of the Division of Radiophysics or that it should be continued for so long; there were other interesting and important projects that could have been pursued instead. The fact that it was pursued so tenaciously is testimony to Bowen's and White's continuing belief in the ability of the program to deliver long-term benefits to the nation. The early work had provided a secure underpinning for all subsequent research by demonstrating beyond all doubt that clouds could be induced to yield rain. What remained to be done was to work out how to induce them to do so in ways that changed the existing, unsatisfactory patterns of rainfall across the country.

Their confidence was evident from the very beginning, in the initial commitment to a research program that even under the best of circumstances was clearly going to take many years to reach fruition, and then in the steady building up of a carefully structured mix of fundamental research on the physics of clouds and practical investigations of cloud seeding. It can also be seen in the orderly development of the investigation from opportunistic single seedings to controlled experiments on single clouds to large-scale experiments over extended areas, and in the highly positive tone adopted in the many public statements that were made about the project. In 1957, in the light of the early apparent success of the Snowy Mountains area experiment, White actually prepared a Cabinet submission proposing the establishment of a Commonwealth Rainmaking Commission that would oversee practical rainmaking work, which he envisaged becoming an ordinary accompaniment of everyday life in Australia, while the CSIRO group continued its long-term experiments.[28] The proposal foundered on the scepticism of Prime Minister Menzies and his reluctance to see the federal government drawn into taking responsibility for drought relief in the states, but those involved remained convinced that routine rainmaking was close to being realised and that appropriate regulatory measures needed to be put in place. CSIRO itself was clearly not set up to handle the increasing number of calls for help from drought-affected communities. Eventually, in the mid-1960s, the various state governments agreed to take responsibility for local cloud seeding operations in response to drought or to increase rainfall in local water catchment areas, with CSIRO acting in an advisory capacity only. This led to the CSIRO group mounting several courses of instruction in the practicalities of rainmaking that attracted both officers from the state governments and significant enrolments from overseas.

Some months before his retirement, Bowen wrote to White's successor as chairman of CSIRO, J. R. Price, concerning the future of the Division over which he had presided for a quarter of a century. He was in no doubt that the future of the radio astronomy program was secure and so focused his discussion of this on the kinds of new (and inevitably expensive) equipment that it was likely to need. When he turned to the cloud physics work, however, he evidently felt it necessary to spell out, one more time, the philosophy that had lain behind it:

> More than half our effort in this field is in basic cloud physics, the remainder going into practical cloud seeding experiments of one form or another. I hold very strongly that the whole reason for a cloud physics programme is the end product — namely the application of the knowledge so gained to augmenting the water supplies of this, the most arid of continents. It would be quite difficult to make a case for basic cloud physics research in the absence of this practical objective. I feel it is of the utmost importance, therefore, that as a matter of policy the practical aspect of the Division's cloud physics work continue to be emphasised, as it has been in the past.[29]

In retirement, Bowen remained convinced that cloud seeding offered long-term benefits to the nation, and his voice continued to be heard. In 1983, two years after CSIRO's cloud seeding program had been shut down, with much of eastern Australia gripped by drought, he argued that the benefits of the program might have been masked by the statistics: the effect of cloud seeding may have been regional rather than local, he suggested, with prolonged increases in rainfall over both target and control areas. The powers that be in CSIRO felt sufficiently threatened that they ordered a major new analysis of the long-term rainfall statistics for the south-eastern part of the continent. This, however, failed to confirm Bowen's hypothesis: 'the study showed no increases in rainfall on either a monthly or a yearly time scale associated with cloud seeding activities above normal variations in precipitation'.[30] Even in the 1990s, a few people continued to argue the inadequacies of the statistics, and that cloud seeding had yielded hidden benefits. By then, however, they were very much in the minority.[31] In Tasmania, however, where the cost-benefit analysis is different because likely clouds occur much more frequently than they do over the mainland, a cloud seeding program has been maintained. There, the Tasmanian Hydroelectric Commission continues to seed clouds regularly in order to maintain water levels in its dams, and there seems to be no doubt that extra rainfall is generated that more than pays for itself. Cloud seeding also continues in various other parts of the world where cloud patterns are more favourable than they are over continental Australia.

Conclusion

CSIRO's cloud seeding research was never cheap. With the benefit of hindsight, it would be very easy to criticise those responsible for persisting for so long with a project that has had so little pay-off for the nation. That would, however, be extremely unfair. There was never any doubt that in this research the stakes were very high. Australia is a very dry continent indeed, and many development projects have foundered for want of water. Moreover, throughout the period since European settlement, the prevailing attitude has always been that we have to overcome nature rather than live within the constraints it imposes. In the post-war world, Australians looked to science to help them do this, and Australian scientists willingly shouldered the responsibility. In the light of the first cloud seeding experiments, and given this general mind-set, it would have been inconceivable, I would suggest, and indeed irresponsible for CSIR not to pursue the rainmaking project. And then, the encouraging results obtained during the early years of the program ensured that the work would continue. The group did a lot of good science on the structure of clouds and ways of modifying them. If there are grounds for criticism, they can only relate to later decisions to keep the project going. But here, too, we must beware of the wisdom of hindsight. To be sure, the first area experiments in the late 1950s were inconclusive, but it does not follow that the project should therefore have been abandoned. On the contrary, the very fact that they were inconclusive indicated that further research was necessary — research that, of its nature, had to run for several years before any conclusions could be drawn. Perhaps some of CSIRO's pronouncements were unduly optimistic, but this, too, only became clear later. A certain amount of optimism is essential in science. How much is too much? Once again, hindsight is not a good basis on which to reach a decision. We also need to remind ourselves that not all scientific research is successful. If it was, there wouldn't be much challenge to it. Perhaps one could argue that the rainmaking group was unlucky in having such extraordinary success with its first seeding experiments, in that this gave rise to unrealistic expectations as to what could be achieved. They were certainly unlucky in that one of the major conclusions that subsequently emerged from the research they did was that not all clouds occurring over continental Australia were potential sources of rain and that, in fact, only a small proportion were. Ultimately, if there is criticism to be made, it should be directed, I believe, not just at those in CSIRO who were responsible for the rainmaking program but also at the Australian people in general whose outlook they shared who, rather than learning to live with Australian nature, simply took it for granted that they could reshape nature on a grand scale for their own ends. Many people, indeed, still refuse to accept this conclusion, which they regard as unduly defeatist. With much of the country once again in the grip of a severe drought in 2002–2004, CSIRO came under renewed pressure to make rain. A new generation of rainmaking enthusiasts has announced schemes for firing silver iodide pellets at passing clouds.

R. W. Home

Rod Home was Professor of History and Philosophy of Science at the University of Melbourne, 1975–2003. He has published extensively on eighteenth-century physics as well as on the history of science in Australia. He edited *Australian Science in the Making* (1988) and has been editor of the journal of the Australian Academy of Science, *Historical Records of Australian Science*, since 1984. He is a Fellow of the Australian Academy of the Humanities.

Hot-air cabinet used in experiments at the Australian Institute of Tropical Medicine in Townsville, in use 1919

NATIONAL MUSEUM OF AUSTRALIA

Maximums and Minimums

The Australian Institute of Tropical Medicine was established in Townsville in 1909. Its first director, Anton Breinl, studied the health risks from heat and tropical diseases. The hot air cabinet, named after the French bacteriologist Louis Pasteur, symbolises some of the hopes and fears of European Australians settling the northern region of the Australian continent. The cabinet was designed for experimental processes requiring controlled heat, but was adapted for a series of tests on wharf labourers in Townsville early last century that measured people's responses to extremes of heat and humidity.

Breinl tested his subjects under controlled conditions, adapting a range of devices such as the cabinet to create hot micro-environments. He measured the respiration, pulse, temperature and fluid loss of his subjects. In one experiment, William J. Young, later a professor of biochemistry, stripped off his clothes and stayed sweating inside the cabinet for 70 minutes.

After exhaustive research, Breinl concluded that labourers were 11 per cent less efficient in hot conditions than in cool ones.

7

Managing climate for the Murray-Darling Basin (1850–2050)

Daniel Connell

The history of European Australian settlers in the Murray-Darling Basin has been shaped by the interaction between three powerful forces: landscape, climate and society. For a long time Euro-Australians in the Basin regarded themselves as victims of the first two of that trilogy but in recent decades, society in and beyond the Basin has grown to be a primal force in its influence on the three-way relationship. Seen from the perspective of the twenty-first century, the relationship between climate and society in the Murray-Darling Basin is interactive and creative, often with unexpected and unwanted results, not passive, dependent or determined.

The first phase of European settlement involved the adaptation of agriculture and pastoralism to climatic conditions very different from those of Britain and North America. Then, early in the twentieth century, there was the reversal of the seasonal patterns of flow along the Basin's rivers, made possible by the construction of large dams. More recently, climate itself is changing as a result of global warming due to carbon emissions from agriculture and pastoralism in the Basin and the world beyond. In addition, there is another complex layer of interaction created by agricultural and economic development over the past 150 years: many natural events — flood, drought, wind, fire and even moderate rainfall — now have impacts that are significantly altered from what they used to be.

In recent decades the unintended environmental consequences of the impact of the first settlers on the landscape and rivers of the Murray-Darling Basin have become increasingly difficult to manage. Consequently, a major theme in the political history of the Basin is the story of attempts to develop institutions that can deal effectively with the changing nature of the interaction between landscape, climate and society.

Dryland farming

New forms of pastoralism and dryland cropping were at the core of the initial adaptation of European settler society to the southern Australian climate. The first major achievement in climatic adjustment was the development of a highly pragmatic and opportunistic system of pastoralism. It evolved rapidly from its early squatting phase and was the basis for Australia's dominant export industry for many decades. The ingredients of success included innovative forms of land tenure, paddocking, disease control, improved pastures, and the breeding of stock suitable to the climate. Important too were technological advances such as wool presses, mechanised shearing, exploitation of groundwater, windmills and refrigeration.

Cropping developed in parallel. From its origins in the mid-north of South Australia in the 1860s and 1870s this new approach to farming brought together technology such as the wheat stripper, mechanical harvesters, the stump jump plough, disease and drought resistant wheats, and integration with international markets.[1] During the closing decades of the nineteenth century these adaptations of pastoralism and dryland cropping merged to create a very distinctive mixed farming system that dominated southern Australia and the Basin for more than a hundred years. In support were state parliaments based on adult male suffrage, able and willing to enact legislation to make cheap land available, build extensive rail networks and foster agricultural innovation. At the district level a complementary system of local government was developed.

These new pastoral and cropping systems were successful adaptations to the climate and landscape of southern Australia. However, a century later it is clear that they are mobilising large volumes of salt previously held inactive in the Basin's sub-soils. As a result of this and other forms of degradation, the landscape of the Basin will provide an increasingly impoverished environment for future generations. In addition, it is now known that these farming systems are depleting the stores of carbon held in soils, thereby threatening long-term productivity and contributing to global warming and the creation of a more hostile climate, an issue discussed later in this chapter.

Making the desert bloom

Nineteenth-century Australian colonial state governments regarded irrigation as the key to unlocking the Australian inland and creating the economic and social conditions needed to foster democracy. In a dry country, the rivers were tantalising. But, although major streams like the Murray were there for all to see, harnessing them was not a simple matter. As the vehicle for populating the inland with democratic, skilled, vibrant and prosperous communities, irrigation needed institutions able to make legally binding decisions about the distribution of water and associated costs and obligations. This required major changes to the water laws, which were based on English common law. Working from a 'riparian rights' foundation, those laws gave no legal support for anyone wanting to draw water from rivers in the volumes needed for irrigation. This discouraged investment and made it difficult for colonial governments to shape development along the lines demanded by their electorates.

The leading figure in the early history of irrigation in Australia was the Victorian, Alfred Deakin. The so-called Deakin Royal Commission of 1885 documented new international developments in irrigation and considered their application to Victoria.[2] In 1886, as the Victorian minister responsible for water resources, Deakin introduced far-reaching legislation which, for the first time in a common law jurisdiction, gave his government control — but not ownership — of water in surface streams. Similar legislation was later passed by other colonial parliaments. Providing a statutory foundation for the water-sharing activities of the new irrigation-based communities was more difficult. After a series of rather painful experiments, Victoria developed an institutional framework for those communities that worked fairly well for most of the twentieth century. Central to these institutions was a balance between public and private investment similar to the new irrigation systems in British India, Italy and Spain but in contrast to those of the United States. The state funded and constructed major headworks such as dams, weirs and regional distribution channels; irrigation communities and property owners were responsible for funding, constructing and operating the system at the local level. To support these initiatives it was necessary for the three states with an interest in the River Murray to manage the river cooperatively. There was good reason for these states to work together and after each was convinced that there was no alternative, that was what they eventually agreed to do.

Reluctant partners

Inter-governmental cooperation in the Murray-Darling Basin was not achieved easily. The devastating 'federation drought' from the mid-1890s to 1902, during which the Murray had stopped flowing for substantial periods of time, showed that cooperation was essential. The 1914 drought, when salinity levels at Morgan reached 10,000 EC, four times higher than humans can drink for short periods, was a reminder for anyone who might have forgotten the earlier lesson. No state wanted to curtail its right to develop as it saw fit, but the disadvantage of this independence was that each drew from a common source, the Murray. Restraint, lack of development, or inability to gain access on the part of one state left more water to be diverted by the other two. Funding for major works such as a future Hume Dam depended on the Commonwealth government within which each state had sufficient influence to block the plans of its rivals if there was not unanimous agreement. In the background, as a result of the confused situation created by the Australian constitution, there lurked the hazards of protracted litigation.

The River Murray Waters Agreement was signed in 1914 and the River Murray Commission established to administer it.[3] The mutual distrust upon which the agreement was based was made clear by the requirement for unanimous agreement for all decisions made by the Commission. That provision, which continues to shape the politics of the Murray-Darling Basin today, gave each state the power of veto over the others if it felt that its interests were threatened.

Desire to make greater use of the water resources of the Murray-Darling system had been a major factor fuelling the campaign for Federation in south-eastern Australia. The representatives elected to the convention of 1897–1898 to draft the Australian constitution came prepared to discuss the issue at length. Reflecting later on the 'rivers debate', New South Wales Premier George Reid wrote that 'schemes of intensive culture for which the river waters were necessary were the only chance we [had] of making our remote interior a hive of human industry'.[4]

Similarly, many years before, Alfred Deakin had asserted for Victoria:

> If Victoria is to continue to progress in the settlement of her people upon the lands and the multiplication of her resources by the conquest of those lands hitherto regarded as worthless, if she is to utilise her abundant natural advantages, bring her productiveness to the highest point, and secure to the agricultural population of her arid districts a permanent prosperity, it must be by irrigation.[5]

The South Australians were also interested in irrigation but it was more the transport potential of the Murray that shaped their thinking. What they all agreed on was that the inland river system could provide great future benefits; the question was: who should reap them?

As soon as they started to plan for irrigation and improved navigation in the late nineteenth century, the states were forced to accept that the annual flow patterns of the Murray and its tributaries were unusually varied in comparison with major river systems on other continents. This unavoidable fact has profoundly affected the institutional framework that has been developed to manage the Basin. The Interstate Royal Commission of 1902, source of many key sections of the 1914 River Murray Waters Agreement, noted the great variability in flow rates. Records quoted in the Royal Commission's report showed the difference between the highest and lowest recorded winter flows of the Wimmera River to be twelve times, of the Goulburn four times, the Murrumbidgee seven times and the Darling ten times.[6] The other difficulty was that the bulk of the annual flows came down the Murray at the wrong time of year. The time of peak flows in the southern rivers was usually winter and spring but the main irrigation growing season when plants needed water was in the late summer and autumn.

The construction of large dams such as Hume reduced the risk of drought and provided a reliable flow that increased the potential for agriculture and habitation along the Murray from Albury to its mouth in South Australia.

STATE LIBRARY OF VICTORIA

To reduce the variation between years and reverse the seasonal pattern of flows within years, the Basin states eventually built a network of dams whose storage capacity, in comparison with mean annual flow, is one of the largest in the world. The accompanying water management system was based on proportions of the available flow rather than defined annual quantities. The River Murray Waters Agreement gave South Australia a fixed entitlement and allowed Victoria and New South Wales to share the rest. During drought, however, South Australia receives a proportion of the flow rather than a fixed amount.[7] The same principles were also applied within states where water is offered at varying levels of security to allow for the differences between wet and dry years. As a further refinement of the system, many irrigators start the season with a low allocation which is increased progressively as the full extent of water availability becomes known.

The level of organisation and coordination required to implement these principles created a system of inter-jurisdictional water management that was more sophisticated than the international norm. A recent study of 145 treaties dealing with inter-jurisdictional river management found that most arrangements are very rudimentary:

> Despite their rich history, a reading of these 145 treaties reveals that the legal management of trans-border rivers is still in its conceptual infancy. More than half of these treaties have no monitoring provisions and, perhaps as a consequence, two thirds do not delineate specific allocations and four-fifths have no enforcement mechanism. Moreover, the treaties that do specify quantities allocate a fixed amount to all riparian nations but one, and that one must then accept the balance of the river flow, regardless of fluctuations. Finally, multilateral basins are, almost without exception, governed by bilateral treaties, precluding the integrated basin management long-advocated by water managers.[8]

In contrast, the institutional structure in place in the Murray-Darling Basin is considerably more integrated. A Basin-wide treaty underpins an extensive network of jointly constructed and operated infrastructure. An elaborate management system is supported by detailed monitoring and methods of enforcement (a combination of the threat of public shaming backed up by the capacity of the Commonwealth government to withdraw much of the funding upon which the states depend). If the state governments had not accepted this degree of cooperation, their capacity to ameliorate the effects of climate would have been severely curtailed.

Despite or because of its long period of gestation, the River Murray Waters Agreement remained in force largely unamended for seventy years, making it possible to transform the Murray Valley into a major agricultural zone whose productivity belied its climate. That did not mean that climate no longer limited human activity in the region. One product of the altered landscape is increased salinisation, a process that will occur at a rising rate for many decades to come.[9]

Salinity

The search to find an effective response to increasing salinity has been the main force for institutional change in the Murray-Darling Basin since the early 1980s. Increasing salinisation of the rivers (and dryland) in the Murray-Darling Basin is the result of changes in the relationship between the Basin's landscape and its climate. It is now known that landscape, climate and society in the Murray-Darling Basin interact over and within a closed system of groundwater basins whose only outlet is the Murray mouth. Understanding of the geophysical processes involved has been recent, the product of the past twenty to thirty years. The confined nature of its hydrological system has made the consequences of unsustainable human activity apparent sooner, and more directly, than would have been the case if it were a region overlying unconfined groundwater systems — as do most parts of Europe and North America.

The Basin's groundwater system was already saturated at the time of European invasion but recharge rates, groundwater levels and stream flows had been stable for a very long time, allowing the landscape and environment it supported to evolve to a high level of complexity. The early generations of European settlers did not understand the dynamics of the Basin's groundwater systems. They did not know that any increase in the recharge rate to groundwater would cause the land surface, and the rivers that drain it, to salinise rapidly.[10] One consequence of the new farming and pastoral systems that transformed rural southern Australia during the late nineteenth century, unnoticed at the time, was that they increased the recharge rate of rainfall to groundwater up to tenfold and often more.[11] This resulted from large-scale clearing of native vegetation and the introduction of new shallow-rooted crops and pastures that are senescent for much of the year.

Even if those changes are reversed immediately, which they will not be, it would take decades and even centuries in some regions for groundwater tables to stabilise once again. As a result, in many parts of the Basin, groundwater systems are rising steadily and mobilising large volumes of salt previously held inactive and harmless in the top few metres of sub-soils. This salt is carried to the land surface by capillary action or moved laterally into streams by rising groundwater systems that now intersect with the base sections of many water bodies. Irrigation development compounds the situation by adding additional water which also carries its own salt load onto these altered landscapes. This creates large volumes of saline water that must drain to somewhere.

Institutional turmoil

The primary task of the River Murray Commission up to the 1970s was the storage and delivery of water to maximise the quantity available for human consumption. However, during the severe drought of 1967–1968, high salinity levels in the lower Murray aroused public concern about water quality. Early analysis suggested that the problem was likely to increase in the future. At first, salinisation was seen as something that required an engineering solution. One such scheme, sometimes known as the Pels scheme, proposed to draw good quality water from the upper reaches of the Murray and pipe it to irrigation areas and towns downstream through a distribution system independent of the river. The middle and lower reaches of the Murray would then be used as a drain to receive saline waste-water from the irrigation areas.[12] The eventual rejection of the Pels scheme and others like it signalled a shift towards a more environmentally orientated approach to river management.

In broad terms, salinisation in the Murray-Darling Basin has two triggers: irrigation and dryland farming. Both activities mobilise large volumes of the salt that has accumulated in the sub-soils over millions of years. During the 1970s and 1980s governments focused on irrigation salinity rather than the salinisation occurring in the dryland farming areas which covered most of the Basin. The effects of irrigation salinity were dramatically obvious and the methods that could control it reasonably clear. Subsequently the salinity impacts of new irrigation projects were brought under control by the Murray-Darling Basin Salinity and Drainage Strategy. Implemented from 1990, it was an inter-jurisdictional pollution credit scheme that significantly lowered salinity in the lower Murray and largely eliminated the destructive salinity spikes that had been occurring with increasing frequency. However, the planners of the Salinity and Drainage Strategy were aware from the beginning that it would only deal with symptoms, not causes.

The complexities of the interaction of landscape, climate and society is severely testing existing institutions in the Murray-Darling Basin. Salinity is just one of many difficult by-products. Some indication of what is needed to deal with salinity is outlined in the Salinity Management Strategy approved by the Murray-Darling Basin Ministerial Council in 2000. Typical of a group of strategies being prepared to implement the Council's Integrated Catchment Management Policy, it will require major changes to land management practices, scientific breakthroughs, innovative product and market development, culture change and new institutions at the regional level. (Although the Integrated Catchment Management Policy has been approved by the Commonwealth and state governments in principle, effective regional catchment bodies are likely to encounter strong opposition from state governments concerned about the prospect of an additional level of government that could bypass them and develop direct funding access to the Commonwealth.)

Implementation of the Integrated Catchment Management Policy will get under way at a time of growing world-wide debate about how best to introduce sustainable natural resources management. One strand of the discussion deals with water. For most of the twentieth century in the Murray-Darling Basin and beyond, decision making about water management was the preserve of a web of institutions that had evolved in the public utilities tradition, a common model for water management in western countries through this period. Those organisations were large, hierarchical, technocratic, culturally monolithic and self contained, but successful according to the criteria of their time.

Their task was to maximise the volume of water available and distribute it, something they did effectively. For a variety of reasons, however, the public utilities were unable to cope with the change in the 1970s and 1980s from maximising water quantity to also protecting and improving water quality. In part this was the product of new ideas about how to fund, organise and operate public institutions. But they also failed to respond effectively to the complexities created by water quality issues such as salinity. These include the need to reform land management practices on private properties in the catchment and work cooperatively with other government organisations, research bodies, environmental groups and assertive catchment communities.

In the mid-1980s, in response to these pressures the River Murray Commission was subsumed into a tripartite structure combining the Murray-Darling Basin Ministerial Council, the Community Advisory Committee and the Murray-Darling Basin Commission. Among other innovations it was agreed that the resources of the river system should be managed to improve biodiversity and sustainability, in addition to promoting production. Advocacy for these competing values was institutionalised, with each jurisdiction sending to the Council and Commission a team of ministers and senior public servants drawn from the agencies representing the different interests. The importance of the community was also recognised although its role was defined very vaguely.

Since it was introduced, this new structure has produced many impressive results, not least, much greater public and government awareness of environmental issues and very high expectations that will be hard to satisfy. However, the environmental condition of the rivers and catchments of the Murray-Darling Basin is still declining[13] and there are growing doubts about the political will of governments to implement the programs they approved in principle as members of the Ministerial Council. Perhaps not surprisingly, some participants and concerned observers involved in the Murray-Darling Basin institutional process are looking to a future major crisis to break the current stalemate.

The future

Crisis, desired or not, could come from a number of issues, most probably from a number of them compounding each other. The drought that gripped much of the Basin in 2002–2004 is one. Previously a notorious source of proposals to build new water storages to drought-proof the country, 'drought' this time round has stimulated public discussion about the future management of rural Australia (not unlike the debate in the 1890s that led to the 1901 New South Wales inquiry into the management of the western part of that state).

The effect on the environment of the changes imposed on the Basin as a result of the development of the last 150 years has been profound. The reversal of the seasonal pattern of flow in most of the major rivers, made possible by large storages and river regulation, has benefited irrigated agriculture, but at a high cost. Combined with the impact of land clearing and other development activities, it has had drastic effects on the flora and fauna that had evolved within the previous environment. This has led to many species becoming endangered and to a significant number of extinctions.

Economic productivity is also at risk. One of the many indicators of the long-term unsustainability of current rural activity is carbon depletion. In Australia, the development of European systems of agriculture has substantially reduced the stores of carbon held in those soils subjected to farming and pastoralism. Recent research indicates that 80 per cent of Australian mineral soils contain less than 80 tonnes of carbon per hectare, representing a loss of up to half the carbon present in the soil profile before European farming began.[14]

Agriculture and pastoralism in the Basin have been and continue to be, a significant contributor to greenhouse gas emissions and climate change, which are predicted to give the region a drier future with more variable weather.[15] How much drier the Basin will become is not clear. However, the authors of a recent report on the subject noted that most irrigation development in the Basin has taken part in the last fifty years when the mean annual rainfall was significantly higher than in the first half of the century.[16] As a result, community notions of 'normality' are probably at the wetter end of the range rather than in the middle. Consequently, the impact of a future reduction compared to the long-term mean may be felt even more sharply.

Managing climate change

So far in the Murray-Darling Basin, discussion about managing the possible impacts of climate change has been pushed to the background by the debate about restoring rivers already degraded under current conditions. There is a strong argument for linking the two: both require water management that is more sensitive to ecological processes and climatic and weather variation. To deal with the degradation that has already occurred, more water needs to be retained in-river, the connections between the main channels of rivers and their floodplains must be revitalised, and a more natural pattern of seasonal flows has to be restored.[17]

To achieve this, the volume of diversions would need to be substantially reduced. Flows would also need to be more responsive to natural weather patterns. If implemented, such a regime would mean higher flows in spring, the traditional time for floods and breeding. During autumn the flows would be lower, slower and warmer. Under present conditions the rivers run fast, deep and cold to deliver large volumes of water to meet irrigation demand. Such a readjustment between the seasons would further reduce the volume of water available for diversion during autumn. In broad terms it is likely that this formula, taken considerably further to adjust for the predicted reduction in available flows, is also what is required to deal with climate change.

The need for these changes to deal with existing degradation was recognised by the five-year review of the cap on diversions in the Basin completed in 2000 and stated more strongly in the literature recently distributed by the Murray-Darling Basin Commission as part of the River Murray environmental flows consultation process.[18] Change in this direction would require irrigators to accept smaller allocations, with an increased proportion of those allocations being low security to allow more flexible, variable environmental management. Despite the demands of powerful irrigation groups,[19] reducing the certainty of water entitlements rather than increasing them is probably what is needed. This is not an argument against redesigning the current methods used to determine water allocations. The case for a system less subject to political and bureaucratic fiat is very strong. From that viewpoint it is reasonable to ask for greater certainty. On the other hand, if the priority is to achieve a sustainable, healthy river the future system should be based on a significantly lower level of diversions and be more sensitive to climate and weather variations. It would not be surprising if the result was more unpredictable, year to year, month to month, day to day, than the current system.

How difficult will it be to increase the capacity for variability?

The institutions that manage water in the Basin are in a good position to respond effectively to the impacts of climate change, but that advantage is in danger of being eroded. Unless diversions from the Murray-Darling system are reduced to a level considerably below any of the options now being discussed,[20] the current push by the Commonwealth government and irrigation industry groups to give greater certainty to diverters is likely to see water allocation systems become less flexible and variable year to year, rather than more. If the level of diversions was fairly low, then certainty could be improved without significantly reducing the natural variability of the system. However, if most of the available flow is promised to diverters at a level that is relatively constant year to year, the volume of water left over for the environment will be very small even in those years that are only mildly dry. This has already happened at the Murray mouth where severe drought conditions, previously experienced once in twenty years, now occur six years in ten.[21]

If reduced capacity to respond to natural climate and weather patterns is the result of the water property debate, Australia will have lost an invaluable, century-old advantage created by the founders of the current system. The fortunate inheritance of the Murray-Darling Basin is highlighted by comparison with other major river systems. The study of 145 treaties dealing with inter-jurisdictional river management, already discussed, found that the practice almost invariably has been to:

> allocate a fixed amount to all riparian nations but one. That one nation must then accept the balance of the river flow, regardless of fluctuations.

Clearly, the last nation in the line, when the overall volume is decreasing because of something like climate change, is not in a good position. The situation of the

Murray-Darling Basin compared with the rivers examined in that study is quite different. As explained earlier, the problem of annual flow variability forced the planners of the River Murray Waters Agreement to opt for proportions of available flow rather than fixed quantity shares. As a result the Basin today is reasonably positioned to deal with the reduced flows predicted to come from climate change. That is not to say that reductions will not cause real difficulty for diverters and the riverine environment. They will, but the system has a process that could manage the predicted change in flows, in contrast to the water management systems in place for many river basins elsewhere in the world.

It seems unavoidable that competition for water in the Murray-Darling Basin will intensify in coming years. In the face of that pressure and given the potential impact of climate change, will the institutions responsible for future water management in the Basin maintain the hard-won advantage derived from their history of dealing with climatic variability, or will that valuable bequest be lost? It is likely that climate change in the Murray-Darling Basin will soon get more attention. Currently, the debate is focused on whether Australia should sign the Kyoto Protocol. However, when the public begins to focus on regional issues, the controversy about how to divide up the cake if it shrinks in response to climate change could become very acrimonious.

A reflection

The interaction of landscape, climate and society in the Murray-Darling Basin is increasingly dominated by society but in no sense can it be said to be under human control. The past 150 years have seen a number of interventions that were initially successful, but damaging to society's interests in the long term. To criticise the original impulse and actions is not appropriate; communities can only be expected to do what seems like a good idea at the time. It is the capacity, or lack of it, to change later in response to new information which is the issue. As the extent of human intervention increases, the cost of not changing to take account of new information is also increasing. The case for climate and weather sensitive, adaptive management able to use new knowledge at the sub-catchment level is now well accepted (even if not widely implemented). Although it was not developed with the environment in mind, the original system of water distribution established early in the twentieth century, with its responsiveness to weather variation between and within years, has some of the features that will be needed if ecologically sensitive, adaptive management is to be applied systematically across the Basin. It is a bequest worth protecting and developing for the future.

Daniel Connell

Daniel Connell is a PhD candidate in the Centre for Resource and Environmental Studies at the Australian National University. He is working on the politics of interstate water management in the Murray-Darling Basin 1850–2050. His publications include *The War at Home*, a social history presenting life in Australia during the Second World War, and *The Confessions of Clyde Cameron*, an interview-based biography of the Labor party stalwart Clyde Cameron, both published by the Australian Broadcasting Commission.

8

The curse of the tropics

David Walker

Tokyo, 1909. Baron Kokura, patriot and high ranking Japanese official, scrutinises a map in his office. It is unclear when the Baron acquired this map, but he is obviously very attached to it and looks at it often. We know it is a spring-loaded map that runs from floor to ceiling, that he has no other map like it and that the sight of the map fully extended or, more precisely perhaps, what the map evokes, draws important speech acts from the Baron. It is a map of Australia, but the Baron is particularly drawn to the Northern Territory. This he refers to as Austral Asia, an unoccupied land of 'magnificent distances' and 'incalculable wealth' and the only part of the globe, according to the Baron, where 'Nature begs and entreats for population'. Kokura is confident that Austral Asia could carry a population of 500 million. He feels its possibilities very keenly and is dismayed that its current occupants treat seductive Austral Asia with such shabby disregard. Northern Australia had no more ardent suitor than the Baron.[1]

The Baron was a figment of the imagination of Francis Hopkins, author of *Reaping the Whirlwind: An Australian Patriotic Drama for Australian People*. The play was published in 1909 and coincided with the growing speculation about the future of Australia's 'empty north'. The Commonwealth parliament debated the future of the Northern Territory at length through 1910, agreeing at the end of that year to transfer responsibility for the Territory from South Australia to the Commonwealth.[2] The concern that Japanese patriots like Baron Kokura, tormented by the need to resolve problems of overcrowding, might turn their attention to northern Australia quickened the tempo of debate. An empty Northern Territory seemed an acute provocation. While it was widely accepted that northern Australia was well-suited to Asian settlement on the grounds that Asian populations were adapted to tropical conditions, white settlement in the tropics was considered particularly perilous, a ghastly narrative of premature deaths, debilitating diseases and decaying Europeans.

Climate and racial character

The close relationship between climate and racial character was one of the foundational beliefs of nineteenth-century science as was the conviction that European settlement in the tropics was fraught with danger. The British historian, Henry Thomas Buckle, was among those who considered climate a primary influence on the growth of civilisations. In his influential *History of Civilization in England*, first published in 1857, Buckle developed an elaborate argument about diet, climate and race.[3] It was part of Buckle's argument that tropical climates encouraged the consumption of fruit whereas cold conditions required large quantities of animal flesh, blubber and oils. It followed that the skills required to hunt, outwit and subdue wild animals for their food were considerably greater than the skills needed to wrestle a banana into submission. It was largely for this reason, no doubt, that Sir Charles Dilke described the banana as the curse of the tropics; it was far too freely available and was not a character-building adversary.[4] Almost a century later, Adolf Hitler was aware of the sinister possibilities of tropical degeneration when he expressed a preference for settling Jews in Central Africa where the climate could be expected to produce weakness and enervation, rather than run the risk that Jews transported to Siberia might become tougher and more resilient.[5]

Buckle went on to make the case that tropical abundance produced populous, low-wage communities, whereas the reverse was the case in Europe. Taking the argument further: the tropics were characterised by despotic regimes where the highest in the land lived in a state of spectacular opulence while the lowest were subjected to grinding poverty and cruel neglect. In tropical civilisations, according to Buckle, the ordinary people exercised no power and their lives counted for nothing. And whereas superstition dominated the thought of tropical Asia, reason found its proper home in Europe.[6]

How Buckle reached his conclusions is less important than his confident finding that tropical conditions produced despotic regimes, cruel tortures and superstitious habits of mind. If white settlement in tropical Australia was to succeed it would have to defy what Buckle regarded as an iron law governing human development in the tropics. The case against the tropics was widespread by the late nineteenth century. The distinguished social scientist, Benjamin Kidd, writing on this question in 1898, declared that the human race developed earliest where the conditions of life were easiest, in the tropics. 'But', he went on, 'throughout the whole period of human history, the development of the race has taken place outwards for the tropics'.[7] The higher civilisations were those at the greatest remove both from tropical origins and tropical conditions. These views were popularised in the 1920s by the Canadian arctic explorer and author of *My Life with the Eskimo*, Vilhjalmur Stefansson. In his examination of climate and civilisation, *The Northward Course of Empire*, Stefansson produced a graph which showed 'the path of supremacy' moving in a rising curve from the heat of Upper Egypt in the ancient world to the colder climates of Europe and North America in the twentieth century. Stefansson concluded that colder climates produced more complex and highly evolved civilisations than tropical ones.[8]

One of the leading academic geographers of the early twentieth century, the American, Ellsworth Huntington, spent a lifetime producing evidence to support the case that cold climates produced higher civilisations than tropical and semi-tropical climates. Huntington's *Civilization and Climate*, published in 1915, argued for the primacy of climate in shaping mental aptitudes and racial capacities.[9] Huntington found many ways of measuring excellence and assigning scores to the climates he liked best. He scored the most favourable climates for human achievement and intellectual development at one hundred. England, being suitably chilly, topped the list. The scores for Australia ranged between the eighties and the thirties, which was hardly reassuring. In relation to Australia, the conclusion to be drawn from Huntington's findings was that British settlers with cold climate aptitudes had raised Australia's civilisational score above the level that could be expected from the climate. It followed that over time Australia's relatively adverse climate would lower the level of civilisation. Australians would become a less accomplished, cruder and more indolent people than the stock from whom they had been drawn. This was a sensitive issue. When the English travel writer, John Foster Fraser, toured Australia in 1909 he thought Australians lacked 'vim'. They showed signs of being drained by their climate. Fraser advised that fresh infusions of immigrants from the old country would be necessary if Australia was to avoid racial decline.[10]

Australia's most prominent geographer of the inter-war years, Professor Griffith Taylor, shared a number of Huntington's concerns about the impacts of tropical conditions on intellectual and physical performance. But climate was also an inseparable dimension of debates about settlement and population, issues on which Taylor was always outspoken and for which he was invariably in hot water. He dismissed the notion that an empty Australia, particularly the empty north, had an almost infinite capacity for population growth. Settlement would be determined by access to markets and relative comfort. The far north of Australia was remote from markets and its climate was uncomfortably enervating, in Taylor's view, and the remote north would remain largely unoccupied so long as there were settlement opportunities in more congenial parts of the continent. Taylor compounded his unpopularity by estimating Australia's optimum population at a figure around thirty to fifty million, a long way short of the hundred million (and more) proposed by population enthusiasts who equated vast open spaces with a vast population. Among them were those who could see no good reason why Australia, being close to America in size, should not also rival America in population, influence and cultural accomplishments. To argue against such a splendid vision of Australia's future seemed unimaginative and unpatriotic.[11]

Limits and opportunities

Australia's often acrimonious debate about climate was as intense as it was because the nature of Australia itself, its limits and possibilities, and its likely place in the world, were still ambiguous. At a time when climate was widely understood to determine the level of civilisation in a society, comments about Australia's climate could have momentous implications, while criticism of the climate could be read as a slur upon the people. Griffith Taylor excited passionate opposition among those who felt that he was simultaneously making Australia smaller and smaller as a field of settlement while also orientalising the continent. In order to establish Australia's defining characteristics, Taylor examined Australia in relation to similar climates across the globe. Taylor believed that by charting homoclimes, regions of similar climate and rainfall, a clearer understanding of Australia's future would emerge. This had the unfortunate effect of appearing to diminish Australia's European connections while confirming that Australia was best understood in relation to Asia. Townsville was linked to Calcutta, Brisbane to Hong Kong and Sydney to Shanghai. Melbourne was given a Southern European cousin, Oporto. Hobart alone approximated England.[12] The geographer had pronounced that Australia was more Asian than European. A large outpost of Empire had been reduced to an appendage of Asia.

The disquieting idea that Australia might be best understood as an Asian landmass was never far removed from debates about climate and population. If Australia was 'Asian' then it appeared to follow that it was destined to carry a population not unlike Java, Japan or China. Anything less might be deemed a failure. It also suggested that Asian settlement would prove more logical and successful than European settlement, raising the question of whether it was appropriate for people not well suited to the climate to claim exclusive possession of the continent in the name of White Australia. In the course of the debate on the future of the Northern Territory a senator commented that the Territory 'presents physical features and climatic conditions which are specially attractive to the Oriental races'.[13] John Foster Fraser, always alert for opportunities to upset Australians, declared the continent an 'Eastern' landmass. Dr Richard Arthur, state member for North Sydney, invited Australians to consider what a dynamic country Australia would be as a colony of Japan. The place 'would hum with industry and wealth production' for the simple reason that Arthur believed the Japanese were better suited to Australian conditions than Europeans.[14] Francis Hopkins was of the same mind, as was Ambrose Pratt, author of a novel of Asian invasion, *The Big Five*, serialised in the *Lone Hand* in 1907. In Pratt's novel, northern Australia has pronounced Asian affinities and a climate that suited 'orientals' better than Europeans.[15]

Percy Trompf, *North Queensland Calls You*, 1933.

PERCY TROMPF TRUST
COURTESY JOSEF LEBOVIC GALLERY

If Australian settlement was a mistaken attempt to transplant Europeans into a climate that did not suit them, it followed that white Australia could expect to have an uncertain future. In fact, the climatic argument was one of the few grounds on which the immigration restriction legislation of 1901 was criticised domestically. In *Colorphobia: An Exposure of the 'White Australia' Fallacy*, published in 1903, the fact that Australia's original inhabitants had black skins was evidence of climatic adaptation. Black skins were better suited to Australia than white ones and the attempt to replace black by white could not hope to succeed. Skin colour, it was argued, 'predicates the possession of powers and abilities that can be put to their best use in certain localities'.[16]

The harmful impacts of tropical conditions on a cold climate people was taken up by an American who had served in the Philippines, Major Charles E. Woodruff, author of *The Effects of Tropical Light on White Men* (1905) and the *Expansion of*

Races (1909). Woodruff maintained that white settlement in Australia was a highly problematic experiment on the grounds that 'climate kills off a race which has migrated too rapidly into a region too light for it'. Life in the tropics, Woodruff maintained, 'was impossible without a dark skin to exclude the fatal actinic rays of light'. Too much exposure to actinic rays produced exhaustion and memory loss, a condition Woodruff referred to as 'Punjab head' because it was so common among the British in India. In Australia, Woodruff was convinced that the hot climate produced a 'nervous temperament' and early exhaustion.[17] Richard Arthur shared Woodruff's conviction that pigmented skin was a necessary protection against the harmful impacts of tropical light.

The degenerative effects of Australia's climate was also a subject dear to the heart of the architect and artist, Hardy Wilson, whose semi-autobiographical novel, *The Dawn of a New Civilisation* (1929), dwelt upon the damage done to Anglo-Saxons in Sydney's semi-tropical climate. It was a familiar tale of exhaustion and diminished creativity. Punjab head had reached Australia. The hero of the novel, who bears a remarkable resemblance to Hardy Wilson, concludes that the attempt to transplant British stock to Sydney would result in failure. Believing that 'the soil of Australia is Oriental' he sought (and found) inspiration in China. Thereafter he took the view that a cultural blending of East and West was the way forward for Australian civilisation.[18]

European settlement in the tropics was the subject of an impressive survey, published by the American Geographical Society in 1939 and written by the Australian geographer, A. Grenfell Price. Price probed every nook and cranny of the debate and was finally unable to give tropical climates a clean bill of health. White settlements in the tropics seemed to make slow progress and the physiological condition of the settlers after sustained exposure to the tropics remained a matter of concern. It was hardly a ringing endorsement for tropical settlement.[19]

An Asian Australia?

It was one thing to argue that Australia's climate might not suit whites and quite another to suggest that whites either relinquish the continent or accept a mixed race community. Those who had misgivings about the climate were likely to have far greater misgivings still about allowing non-Europeans to settle in Australia. Where climate caused unease about degeneration, racial mixing was thought to guarantee a degenerate community. In the case of races deemed strong and enduring, and the Chinese were the pre-eminent case, there was the added concern that the law of racial incompatibility would ensure that one race would seek to eliminate the other. If the Chinese gained a foothold in Australia it was considered all too likely that they would supplant Europeans in the battle for racial supremacy.[20] In short, it might be conceded that Australia's climate was not ideal for white settlement but all the alternatives to a rigorously maintained white Australia seemed far worse. If Australians were not prepared to move aside they had better make the most of their continent, tropical climates and all. This was a message Billy Hughes was fond of repeating: 'Our critics tell us that if we persist with the "White Australia" policy we may lose the country. Perhaps they are right. We certainly will lose it if we abandon that policy'.[21] This was the same Hughes who freely acknowledged that tropical climates 'burned the strength out of them [the British] and withered away their blood with fevers and exhaustion'.[22]

The determination to uphold white Australia produced a body of climatic patriots of whom there was no greater example than the ostentatiously vigorous Randolph Bedford, writer, entrepreneur, politician and white Australia zealot. The Australian sun had no firmer supporter. Bedford was born in Sydney in 1868 and grew up in a world in which the nature of the Australian continent was still being determined. Bedford soon became convinced that Australia could hardly have a bright future if the climate was considered in any way oriental and degenerative. He denounced anyone, professors in particular, who talked loosely about the similarities between tropical Asia and Australia's tropical north. They were as different as chalk and cheese: oriental heat was certainly enervating and degenerative, but Bedford declared that Australian heat was energising: 'There's no enervation as in Java or Singapore. It's Australian, not tropical heat'.[23] Australian heat made bolder and more adventurous types and more resilient bodies; it was entirely therapeutic and the more you got of it the better off you were. He developed the habit of looking into the sun in order to draw down its powers. In his travel book, *Explorations in Civilisation*, Bedford came out as an unequivocal supporter of heat: 'I hate cold and snow as I hate death — and cold is death. It is the people with the greatest capacity for enjoyment who love the sun — the man who hates heat is half-dead already'.[24]

Bedford was no less dismissive of weak-minded talk about Australian deserts as if there was something Saharan and mysteriously Middle Eastern about central Australia. Desert landscapes had become one of the classic formulations of orientalism in the wake of the French occupation of Egypt, made famous in nineteenth-century art through the paintings of Delacroix and in literature through Flaubert's writings.[25] Shimmering mirages, harems and despotic rule had no place in Bedford's understanding of Australia's future as a hot-climate country. He acknowledged no region in Australia so dry that irrigation and settlement could not redeem it. He was a great believer in rain following the plough.[26] Australia was truly a land of opportunity designed for people imbued with energy and colonising zeal.

A climate of redemption

Redeeming the Australian climate from the negative associations of tropicality was an important dimension of the nationalist project. Australians had to be encouraged to love their sunburnt country, a mission the monthly journal, the *Lone Hand*, undertook with enthusiasm. The *Lone Hand*, an offshoot of the *Bulletin*, took a Bedfordian view of Australian possibilities, promoting bounteous Australia at every opportunity. The benefits of Australia's climate were considered important enough to form part of the journal's rationale. It wanted to 'teach Australians how to live in accordance with the conditions of their own sunny clime and not according to the dour and depressing customs illogically imported from chilly distant regions ...'[27] The link between sunshine and youthful optimism seemed perfectly suited to the forward-looking spirit of young Australia, the first new nation of the twentieth century. The first editor of the *Lone Hand*, Frank Fox, went a step further and dedicated his illustrated survey of the Commonwealth, published under the title, *Australia*, in 1910 'To the fierce sun of Australia, which tempers men as fire tempers steel; to the gracious sun of Australia, which makes nature teem with bounty; to the glowing sun of Australia, which warms the heart, enkindles the eye, ruddies the cheek, this in the tribute'.[28]

The celebration of sunny Australia was also connected to the progressivist and vitalist strain in post-Federation Australia. Australian progressives, as Michael Roe has shown, were keen to establish the conditions for a new beginning in Australia.[29] The promotion of health, vitality and racial renewal were fundamental progressive objectives and in each case it was understood that science would help overcome the limitations of the past. Humanity would be led towards a brighter future, a new dawn, outgrowing the oppressive weight of the past, its sickness and defeatism. Could there be a better place than a new continent for such a fresh beginning? Allusions to sunny Australia encouraged the hope that Anglo-Saxons could look forward to the renewal of the race and a brighter future in Australia. As for tropical Australia, Warwick Anderson has shown that science was expected to solve the problems attributed to tropical settlement.[30] A young, progressive and scientifically aware nation, rather than succumb to negativity and gloom, would make tropical settlement safe for Europeans. Australia would lead the world, thereby fulfilling, as the *Bulletin* saw it, 'our glorious destiny as the model white tropical state of the globe and of history'.[31]

How far the discovery of warmth and the possibility of pursuing more physically active lives might have been aided by a rejection of the heavy décor and darkness increasingly attributed to the departed Victorian era is an open question, but Edwardian modernity certainly embraced lighter clothing and encouraged a new freedom of movement. From the 1890s women had taken to the bicycle and the open road in increasing numbers and while there was anxious commentary about the health impacts of cycling, the realisation of new freedoms was apparent.[32] The Australian climate promoted the outdoor life and brought a closer contact with nature. It promoted self-renewal.

The *Lone Hand* view that the English climate produced 'dour and depressing customs' had echoes in the writings of Hippolyte Taine and Emile Boutmy. Boutmy attributed a certain vigour and imagination to England's 'grey and misty climate', but he also blamed the 'morose' temperament of the English on their climate. By contrast, the limpid air and brilliant light of France produced clarity of mind and a taste for abstract ideas unknown to the English.[33] A. G. Stephens, throughout his career as a critic and most influentially as editor of the *Bulletin's* Red Page, drew similar connections between climate, physiology and cultural expression. Stephens was also in love with France and preferred its climate to England's, just as Bedford preferred Italy to England. Between tropicality and bracing cold lay the pleasing prospect of Australia developing a Mediterranean lifestyle and a temperament to match.[34] In 1908, Charles Daley of Geelong, Victoria, writing in *Australia To-Day*, a handsome publication dedicated to the promotion of Australia's future, declared that 'the warm Australian sun has got into the blood, kindling a quicker and more mercurial temperament … with a tendency to take a brighter and less serious view of life'.[35] It seemed no bad thing for Australians to be brighter and less serious than their forebears.

Australia's Mediterranean future received extensive treatment in the writings of Philip Muskett, one of the most prolific writers on diet and health in late nineteenth and early twentieth century Australia. He declared Australia to be 'practically Southern Europe' and as a consequence argued for a dietary revolution in Australia away from the reliance on three meat meals a day and towards a diet favouring vegetables, fruit, fish and wine.[36] A change of diet, Muskett maintained, would also bring what he considered morally virtuous industries like wine growing, fishing and market gardening into greater prominence. But there was also a vision of outdoor elegance, of parasols and sunshine, wine and fruit in an Australia that had learnt how to enjoy warmth and the outdoor life.

Where the virtues of a warmer climate were extensively reviewed before the First World War, warmth in all its meanings appeared much more desirable in the aftermath of the war. As Paul Fussell has argued, mass death had become so closely associated with frozen trenches and the brutal winter of 1917 that the celebration of bracing cold appeared as illogical as the war itself.[37] The very spirit of Europe seemed to have frozen into ugly hatreds and animosities. Through the 1920s travel writers escaping memories of the war set off to find warmer climates where a renewal of the spirit and the rediscovery of pleasure seemed more assured. The Sydney writer, Arthur Adams, sensing the new mood, published *The Australians*, a novel in praise of the Australian climate. It was one long celebration of golden beaches, surf and sun-bronzed bodies. Indeed, there was hardly a more frequently used description of Australians in the inter-war years than 'sun-bronzed', though 'virile' was another favourite. Adams, who loved both terms, wrote in the *Bulletin* and *Lone Hand* tradition of climatic patriotism. Like Charles Daley of Geelong, he thought that the sun had entered the blood of Australians, transforming them in the process into a bold, joyous, adventurous race.[38]

The American geographer, Ellsworth Huntington, made his first trip to Australia in 1923 to attend the Pan Pacific Science Congress and was rather taken aback by the number of people who told him that 'Australia had the finest climate in the world'. He wondered if it was a 'patriotic duty' among Australians to say nothing derogatory about their climate.[39] Huntington may also have seen a frontal assault on Griffith Taylor in *Smith's Weekly*, accusing him of falling victim to the American fad of climatology. *Smith's Weekly* laughed off the weighty Huntingtonian studies as 'special pleading for pneumonia and blizzards'.[40] Since criticism of Australia's climate was often accompanied by a thinly veiled attack on the sustainability of white Australia it is reasonable to conclude that by the 1920s defending the climate was certainly considered a patriotic duty, and especially so if the critic was an overseas professor looking for evidence of degeneration and Punjab head. Wherever he travelled in Queensland, Huntington was on the look-out for the slovenly behaviour, mental confusion or irritability that told him that tropical heat was having its expected impacts on the locals. He left Australia convinced that his theories were being borne out.

The view from tropical medicine

Medical opinion in Australia had already moved away from this position. The subject of tropical settlement had been before the Australasian Medical Congress since 1911 when it resolved to investigate 'the possibility of the permanent occupation of tropical Australia by a healthy, indigenous white race ... ' Delayed by the war, the report was not published until 1920. It provided a cautiously worded endorsement of permanent settlement, concluding that there were no 'inherent or insuperable obstacles' to white settlement in the tropics.[41] The report found no convincing evidence that tropical conditions produced permanent physiological damage or racial deterioration, although it was agreed that every effort should be made to make tropical settlement a science. Diet, clothing, housing, conditions of work and tropical diseases would all be subjected to the closest scientific scrutiny. Tropical settlement in Australia would be a closely monitored and carefully managed experiment operating at the cutting edge of modern science. Throwing caution to the winds, Arthur Adams declared Australia to be 'the largest-sized and the most tremendous experiment ever tried in race-building, and Australia knows it'.[42] The same defiant, world-beating rhetoric pervaded one of the glossiest and most lavishly illustrated books published in 1918, E. J. Brady's *Australia Unlimited*, another tribute to the best climate in the world.[43]

In the mid-1920s, Raphael Cilento, director of the Institute of Tropical Medicine in Townsville, supported the case for tropical settlement in *The White Man in the Tropics: With Especial Reference to Australia and its Dependencies.* Cilento dismissed the view that tropical heat was a serious problem in Australia, arguing that the key to healthy settlement lay in controlling disease. This was deemed to be less of a problem in Australia than tropical Asia because the Indigenous population was small. Cilento clearly regarded this as a good thing and a smaller Indigenous population, which he confidently anticipated, an even better one. Cilento found no barriers to tropical settlement in Australia that modern medical science could not overcome. Sounding all the familiar notes of tropical triumphalism that characterised nationalist rhetoric in the 1920s, Cilento celebrated the tropical settlers of Australia's north as 'the largest mass of a population settled in any part of the tropical world, and represent a huge, unconscious experiment in acclimatization ...'[44] Australians were encouraged to believe that a great experiment in race-building was under way and that the eyes of the world were on northern Australia.

The idea that over time Australians would become an enervated, prematurely exhausted people met fierce resistance. Such views presented an unacceptable challenge to a white Australia. Only a healthy, energetic people could hope to subdue and settle an entire continent. The hardiness of the pioneers and the daring of the Anzacs were called upon to serve the cause of virile nationhood. Australians, it was argued, were developing into a sturdy, sun-bronzed race in which athletic prowess and a love of the outdoors were the norm. The often strident affirmation of the Australian as a figure of exemplary energy through the 1920s was accompanied by an insistence that Australia was uniquely blessed by a perfect climate that guaranteed warmth and renewal.

A continent of challenges

The re-fashioning of the white Australian as the ideal settler served other purposes. By making Australia a continent of unique challenges to settlement, what had once been regarded as a mighty space on the map, Kokura's Austral-Asia, was transformed into a much more particular place requiring local knowledge and specific adaptive skills. Settling Australia was no longer a matter of decanting millions of Japanese into the outback, as Richard Arthur had proposed, in the belief that they would automatically make more successful settlers than white Australians. Far from offering attractive settlement prospects for Asia's millions, Australia was increasingly represented through the 1930s as a harsh, dry and inhospitable land.

Rather than dwelling on the failures of white settlement, there was a growing emphasis on the hard-won successes of the pioneers, on their grit, wry humour and capacity to endure. It now seemed important to talk up Australian achievements in the field of settlement and development while at the same time minimising Australia's appeal to immigrants from Asia.[45]

Australia's temperate climate, as Warwick Anderson has shown, was progressively cleared of all the serious charges levelled against it up to the 1870s and much the same thing happened with tropical climates through the twentieth century.[46] With the decolonisation of Asia through the 1940s, arguments that attributed lethargy and lack of enterprise among the 'natives' to their hostile climate and superstitious beliefs gave way to socio-economic analyses of under-development and the impact of colonisation. In America, Huntington's climatically driven explanations of civilisation and behaviour lost favour from the 1930s onwards and his monumental labours were largely discontinued.

In Australia, the belief that successful tropical settlement provided further proof of Australian adaptability and ingenuity was maintained into the post-war period. Raphael Cilento's co-authored history of Queensland was published in the late 1950s as *Triumph in the Tropics*, a title that harked back to the defiant rhetoric of the 1920s.[47] But as Cilento looked back to the great battles of the past, tropical climates were increasingly seen, or marketed, as sensually rich and relaxing playgrounds for holidaying adults. They promised renewal. Where Tasmania offered nineteenth and early twentieth century travellers a bracing climate that would renew the blood and sharpen the intellect, from the 1950s Queensland increasingly emerged as the preferred home of the relaxing classes. The state's rehabilitation was certainly assisted by air conditioning and improved refrigeration, but Bedford's passion for sun and heat had won the day. Tropical climates were celebrated as an ideal respite from the daily burdens and stresses of modern, urban society: the tropics encouraged a return to playfulness and pleasure, colour and music. In this utopia of leisure, tropical abundance was something to celebrate. The banana, so cruelly maligned by Sir Charles Dilke, joined its tropical friends, the pineapple and the mango, in a demonstration of just how satisfying the good life could be in the tropics.

David Walker

David Walker is Professor of Australian Studies at Deakin University and held the Monash Chair of Australian Studies at Georgetown University, Washington DC, in 1997–1998. He has written extensively on Australian social and cultural history. His publications include *Anxious Nation: Australia and the Rise of Asia, 1850–1939*, University of Queensland Press 1999, which examines Australian responses to Asia, including debates on population, race, climate and tropicality. *Anxious Nation* was awarded the Ernest Scott prize for the best history of Australia or New Zealand published in 1999–2000. He is a Fellow of the Academy of Social Sciences in Australia.

Solar car *BP Quiet Achiever*, 1982

NATIONAL MUSEUM OF AUSTRALIA
PHOTOGRAPH BY GEORGE SERRAS

Sunny Periods

On 7 January 1983 Hans Tholstrup and Larry Perkins arrived at the steps of the Sydney Opera House, having driven from Perth in a car powered completely by solar power. This vehicle, named the *Quiet Achiever*, averaged only 23 kilometres per hour on its 20-day journey, but it brought the possibility of solar cars to life. In 1987, Tholstrup organised the first World Solar Challenge in Australia. Solar vehicles normally use a battery pack but the 3000-kilometre World Solar Challenge allows no recharging once the cars have left Darwin, so the speed of this event is a reasonable measure of what is possible using solar power. The highest average speed achieved in this event over one day was 94.21 kilometres per hour by the *Honda Dream* in 1996.

9

'All just melted with heat'
mothers, babies and 'hot winds' in colonial Melbourne

Janet McCalman

On 6 March 1857 Isabella Wyly wrote feelingly from Melbourne to her sister in Ireland:

> You say I have told you nothing about the Climate, but what with dust, & Heat & hot winds & Flys & an Insect that the call Moskitoes we do not know what to do with ourselves just now. We ar suffering very much from the hot weather. It has lasted just 7 or 8 weeks, which causes much illness, and [at] present we ar by the Thermometer 90 degrees in the shade. So you may fancy from that that what it is but you cold scarsley believe what it is. The old colonists say it is much cooler now than it was 17 or 18 years ago, but I cannot think that people could live if it was much hotter. It is very pleasant in winter time, unless some times we have very heavy falls of rain, which makes our roads very unpleasant for walking for the ar not paved as are atome. We have had no shower of rain for the last 12 weeks, but taking it all together the say it is a very Helthy Place. But for the children it is mos difficult to bring them up. It is very trying for them. It makes me quite cold to think of winter [?] yet for just now ware at all just Melted with heat.[1]

Most colonial immigrants to the south-eastern seaboard of Australia delighted in the benign freshness of the climate, especially in the winter. Their phthisical chests improved and their winter bronchitis was less severe. As Isabella testified, however, the summers could be nasty, especially for children. Indeed in the antipodes, opposites were true in infant death as in so many things.

Infant mortality was an acknowledged scandal from the beginning of colonial settlement. To some extent, the scandal was that it existed at all in this pristine European settlement. Doctors were concerned about it, but also contested each other's estimates, some of them mindful of any bad publicity back in Britain that might deter respectable new settlers. Warwick Anderson has written of these early debates over summer heat and European constitutions as part of a much longer medical discourse around climate and health that persisted well into the twentieth century, which was concerned with the 'cultivation of whiteness' and the scientific framing of racial destiny.[2]

The hazards of summer

Historians of motherhood and infant care have concentrated on the environmental hazards of hand feeding, impure cow's milk, contaminated water and living spaces, to explain the annual 'slaughter of the innocents' that afflicted Australian communities each summer. We have been warned not to heed contemporary doctors' criticism of maternal ignorance and/or negligence, and indeed much of the historiography of motherhood has implied that expert knowledge was enforced, especially in the first half of the twentieth century, at the expense of women's traditional knowledge and domestic autonomy. If we have been urged to read sources such as doctors' advice on infant care 'against the grain', it might well be worthwhile to look again at their face value now that we have learned to be more sceptical. If medical theory before the wider acceptance of germ theory and scientific method is of historical rather than practical interest, the empirical knowledge acquired over time by medical practitioners can perhaps be re-examined for its usefulness.

Dr J. William Mackenna alarmed many in 1858 with claims that, in the summer months, births scarcely exceeded deaths in Melbourne and that an unacceptable toll of infant life was being exacted by the extremes of burning heat and sudden cold in southern Victoria. Analysing the deaths of children under three years of age in the registration district of Melbourne, he noted that just sixty died in June and July 1857, whereas 163 died in the following December and January. Deaths from dysentery, convulsions and teething generally indicated diarrhoea, but he looked to an external environmental primary cause — 'the prolonged, dry, and scorching Heats' — to explain this 'aggravated mortality'.[3] His writings aroused public anxiety and the *Australian Medical Journal* felt obliged to respond with a withering critique of Dr Mackenna's statistical technique: maintaining that rather than eight out of ten children born in Melbourne dying before their first birthday, indeed infant survival was better there than in every registration district of England.[4]

Mackenna was not entirely mistaken however, given his evidence was significantly anecdotal. Our research now suggests that infant death was greatly under-reported, even as late as the 1880s, in particular by the poor and disorganised, whose babies were at greatest risk. In fact, our attempts to trace the fate of all the babies born at the Lying-in Hospital (now the Royal Women's Hospital) between 1857 and 1900, shows that a seasonal mortality rate of almost eighty in every hundred births does not seem all that far-fetched, especially in the gold rush period.[5] And as we examine the death certificates of those we do find, causes of death appear more complex than many commentators have since allowed.

Mothering in poverty and filth

More reliable infant mortality figures for Victoria begin in the 1870s, although too many small and presumably weakly babies of unmarried mothers delivering in the Lying-in Hospital disappear from the records without trace for one to be particularly sanguine about their survival. While institutional deaths were recorded and more conspicuous, entrepreneurial nurses were no doubt conscious of officialdom's interest in their practices and were compliant, less reliable were the casual nurses who agreed to hand raise a mother's baby and whose negligence might disappear down the cesspit rather than be reported to the registrar of births and deaths. In January

1875, for instance, Susan McGrath was arrested in Richmond for being drunk and disorderly. In her care was found a young baby who soon after died from starvation. McGrath stated that

> she took charge of it on 21st ultimo in Hoddle Street, under agreement to nurse it at a rate of 8/- a week from a woman who stated that she resided off Hoddle Street, Collingwood, and who is described as Scotch, 40 to 45 years of age, very short and swarthy complexion, dark hair, upper front teeth wanting; dressed in dark dress, old Paisley shawl, black lace bonnet; stated she had a family of her own and she was nursing the child for a young woman who was in service up in the country.[6]

Another instructive case was of Naomi Patching, born in the hospital in October 1874, after a long labour that had to be assisted by forceps under chloroform.[7] Four months later, on 26 February, this once 8 pound, 20 inch [3½ kg, 50 cm] infant was dead from 'marasmus' or 'wasting' suffered since birth.[8]

It was towards the end of what had been a ferocious summer, with already nineteen days over 30°C in December, January and February, and two more heatwaves to follow in March (see Table 1). The worst had been from 18 to 22 January, when the temperature exceeded 41° for three consecutive days, peaking at 43.5° on the 20th.[9] On 19 January, William Joseph O'Leary succumbed at five months from softening of the stomach and convulsions: a condition the Lying-in Hospital doctors often associated with heavy drinking by the mother and the giving of alcoholic stimulants and 'improper feeding' of babies.[10]

On 22 January, the third day over 40°, Mary Ellen Ryan died in Curzon Street, North Melbourne from 'congestion of the brain', after three days of illness: heat stress mimicking the feverish state of meningitis. Her 27-year-old mother, Julia, was an unmarried Irishwoman, and her friend who took the child's body for the registration of death was illiterate. Julia was to lose her next two babies likewise in summer; she stayed in North Melbourne, gravitating by 1888 to the rougher area in the east.[11] None of her children survived the first six months of life.

On the same day another Irishwoman's baby died, at Napier Street, Fitzroy: John Mortimer, the son of a shearer, died after suffering diarrhoea and convulsions for six days.[12] And two days later Thomas Wareing, the second child of an Englishwoman and her contractor husband, died at eleven months after two months of diarrhoea from dentition. His mother was to have ten live births in the next fourteen years, the last at the age of forty-eight. Four of the children died in infancy, two daughters in late adolescence, and just three of the original ten were traced to have reached a full life span. Mrs Wareing herself lived until 1920; her husband had died relatively early at the age of forty-one.[13] This level of reproductive attrition was not uncommon, and while some women seemed to lose all their babies before they reached six months, for married women there were many whose children died at many ages and whose young adult offspring remained vulnerable to accidental death, tuberculosis and death related to child-bearing.

Which brings us back to Naomi Patching. Naomi should not have died from marasmus. She was a big, full-term baby. She was her mother's fourth, although records can only be found for two before she was born. She was registered as unnamed

TABLE 1 MELBOURNE SUMMERS AND EXTREME EVENTS[1]

Year	Month	Av. Max.	Max.	Days over 30°C	Heatwaves 3 days or more	* days 40°+ # heat waves ≥5 days
1869-1870	Dec	26	42.4	8	2	1*
	Jan	26.8	41.7	11	4	0**
	Feb	26.5	42.7	8	1	0*
1870-1871	Dec	23.7	36.3	6	2	0
	Jan	26.8	36.7	8	2	0
	Feb	25.3	34.3	6	2	0
1871-1872	Dec	25.9	41.1	7	2	0*
	Jan	27.1	39.6	9	1	1#
	Feb	24.3	35.5	6	1	0#
1872-1873	Dec	24.0	32.3	2	0	0
	Jan	24.9	38.3	7	2	2
	Feb	25.5	39.1	5	2	0
1873-1874	Dec	27.3	38.4	11	2	2#
	Jan	25.5	37.7	8	3	1
	Feb	25.1	38.3	7	1	1
1874-1875	Dec	24.1	39.3	4	0	0
	Jan	26.9	43.5	7	1	1#***
	Feb	24.9	37.9	8	2	0
1875-1876	Dec	22.4	34.0	4	1	1
	Jan	25.7	39.3	9	2	2
	Feb	26.1	39.3	8	2	1#
1876-1877	Dec	25.1	43.7	6	2	0*
	Jan	23.8	38.1	6	1	0
	Feb	25.9	37.9	8	2	1#
1877-1878	Dec	24.1	36.0	8	1	0
	Jan	25.7	37.3	7	3	0
	Feb	24.2	32.9	5	2	1#
1878-1879	Dec	24.3	39.7	5	1	1
	Jan	25.3	41.1	8	2	1*
	Feb	25.9	40.0	8	3	1*
1879-1880	Dec	22.9	36.7	4	1	0
	Jan	25.2	41.3	9	2	1*
	Feb	27.4	35.7	6	2	1
1880-1881	Dec	23.3	39.9	2	0	0
	Jan	26.2	37.7	9	2	1
	Feb	25.1	37.3	7	2	0
1881-1882	Dec	22.5	36.1	6	2	0
	Jan	25.9	43.6	9	3	0 ***
	Feb	26.2	40.6	6	2	0 *
1882-1883	Dec	24.0	35.7	6	1	0
	Jan	26.4	40.5	8	2	1**
	Feb	25.7	37.7	6	2	1
1883-1884	Dec	23.7	37.3	5	1	0
	Jan	23.3	36.5	4	1	0
	Feb	23.5	38.1	4	1	0
1884-1885	Dec	22.5	31.8	4	1	0
	Jan	24.8	36.9	4	0	0
	Feb	24.9	38.7	6	0	0
1885-1886	Dec	24.2	35.1	7	2	1
	Jan	25.7	37.7	9	3	0
	Feb	22.4	40.1	3	1	1*
1886-1887	Dec	23.0	36.0	5	2	0
	Jan	27.0	40.5	10	2	1*#
	Feb	25.8	37.5	9	2	1#
1887-1888	Dec	24.6	34.7	8	3	1
	Jan	23.2	40.0	4	0	0*
	Feb	24.7	36.2	6	2	0
1888-1889	Dec	25.9	35.7	7	2	0
	Jan	25.3	37.3	7	2	0
	Feb	24.5	35.5	6	1	0
1889-1890	Dec	23.7	37.7	5	1	0
	Jan	28.8	38.6	14	4	2
	Feb	27.8	39.7	9	1	1#

[1] Heatwaves are defined as days when the temperature reached at least 30°. The extreme events are then divided into those that lasted for three or more days, * denotes a day over 40°, and # those that lasted five days or longer.

— perhaps her mother was having trouble recovering from her forceps delivery and was depressed and uninterested in her. Certainly, any baby born from a long labour with complications was more at risk of later infant death than a baby born after an uneventful labour, as arduous labours and surgical interventions damaged the mother–baby bond as a couple — damage that ultimately proved fatal for the baby, no matter how robust at birth. But Naomi's family had problems. Only two of her at least seven siblings survived childhood, and then almost despite their family rather than because of it. The next child, Mary, for instance, was born at the hospital in 1876. Four years later she was found wandering about the city streets, by a Lady Visitor for the St Paul's District. Her father was now thought to be dead; her mother lived in one of the roughest alleys in the city. Mary was committed as a neglected child, went into foster care and by 1890 was released on probation in Ballarat East. Seven years later, she married a widower, then lost two young babies before successfully rearing a small family. She lived until the age of fifty-six. Her only surviving sister married at twenty, was widowed at twenty-four, remarried then died in Port Melbourne at the age of thirty-one. Mrs Patching disappeared from the historical record.[14]

These case histories are worth detailed examination because they were in fact typical of the stories of babies born in the Lying-in Hospital in the 1870s and 1880s. The mortality of these babies of the inner-city poor at times approached 70 per cent in the mid-1870s and even 1880s. And what is puzzling about their case histories is that the medical diagnosis of cause of death often does not conform to the classic signs of gastro-enteritis or gut infections. In other words, the received interpretation of Australian seasonal infant mortality fluctuations does not quite fit. It may be that the doctors were careless and wrote 'diarrhoea' when they should also have included 'vomiting', but in fact they were quite careful about their nosologies and Victoria's system of death registration was one of the best in the world. The difficulty is that diarrhoea is an indicator of many different problems in babies and young children, disturbances to the flora of the gut that may not have their only origin in an infection. Sudden catastrophic gastro-enteritis in the young is most commonly caused by toxins, typically those generated by *E. coli* contamination, rather than by reproducing bacteria. Indeed, toxins from *E. coli* makes sense of the diarrhoea of dentition — when children were crawling, putting dirty fingers in their mouths, and in nineteenth-century Melbourne, playing amongst an astonishingly high level of environmental contamination by human and animal faeces.

The stench of inner Melbourne cannot be exaggerated. Not only had the city grown rapidly — one of the great 'instant cities' of the New World that stretched *ad hoc* infrastructures to the limit — but it had some intrinsic environmental disadvantages. Most of the city was perched on a thin layer of topsoil over a base of clay. In the tight spaces of the residential areas of the inner city, classic European cesspits failed to drain waste into the subsoil. Only in the sand belt suburbs could human waste be trusted to be absorbed into the ground and many householders there simply emptied chamber pots into the garden. Elsewhere, cesspits had to be emptied or they filled with rainwater and overflowed over gardens, down back lanes and under houses. A thick, black, putrid sludge filled every nook, hole and channel in wet weather.[15] As the Board of Health for the City of Melbourne reported in

1876, after two years of epidemics of measles and streptococcal infections — scarlet fever, erysipelas and puerperal fever:

> The contagium which causes epidemics is the offspring of insanitary habits, and is nurtured and spread by the impure air of unventilated houses. The ground upon which the town stands is saturated with subsoil water, which must be drained away by underground pipes before we can expect to be healthy. In a great many parts, the houses are not only damp, but impregnated with poisonous gas from pent-up sewage.[16]

Hot spells

It was easy then and easy now to blame environmental pollution for the excessive rates of diarrhoeal disease in colonial Melbourne. If the miasmatic theories of the time were limited, they none the less correctly identified noisome poor sanitation as an indicative public health evil. Deaths from typhoid and infant mortality fell steadily from the mid-1890s — and especially after Melbourne began to be sewered from the turn of the century. But infant mortality also fell in most parts of Australia, some of which, like Brisbane, were not sewered until well into the twentieth century. Indeed the other index that falls in parallel with the infant mortality rate is the birth rate, suggesting the significance of concurrent changes in family life, except for some significant divergences when infant mortality suddenly spiked.

The spikes correspond in New South Wales to the El Niño events of 1877, 1884, 1888, 1897, 1899, 1902 and 1915.[17] In Victoria they correspond with the most severe summers, which tended to occur just before New South Wales' worst summers. The difference in severity was not just in mean temperatures, but also in the frequency and length of hot spells over 30°. When there was less cooling between hot spells, the young and the weak had less chance to recover before the next onslaught. The sharper spikes in Victoria also reflect a newer, more dislocated society undergoing

GRAPH 1
INFANT MORTALITY RATES EXCEEDED BIRTH RATES IN NEW SOUTH WALES AND VICTORIA, 1870–1930[1]

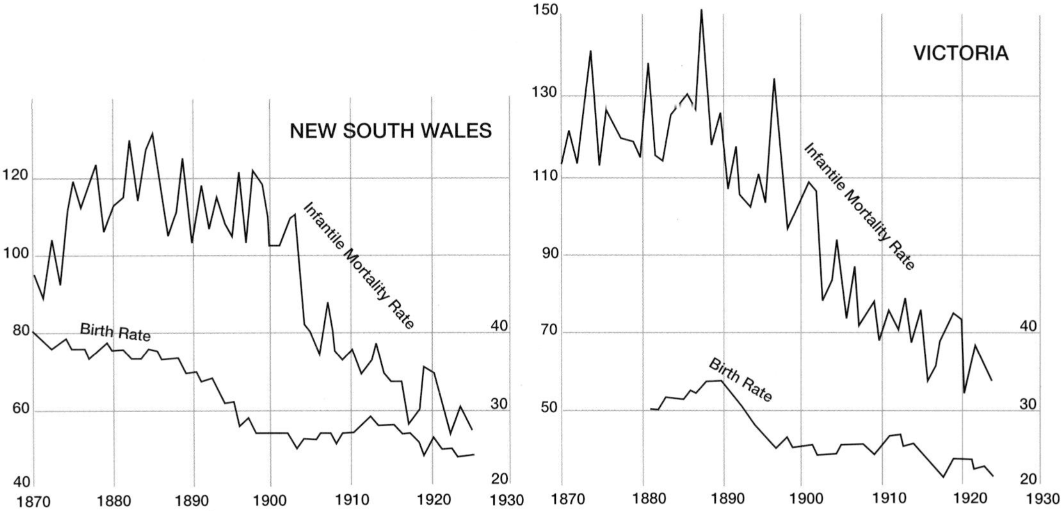

[1] From J. H. L. Cumpston, *Health and Disease in Australia*, ed. M. J. Lewis, AGPS, Canberra, 1989, p.108

rapid urbanisation in the 1870s and 1880s, with an instability of living conditions. Ruth Morley and I have interpreted the fall in mean birth weights in the Lying-in Hospital during the 1870s as reflecting a complex of factors including selection policies of the hospital, life chances of young colonial women and emigrants, structural economic change and falling urban food prices during agricultural slumps.[18]

GRAPH 2
MEAN BIRTH WEIGHT OF BABIES BORN TO MOTHERS BORN IN AUSTRALIA, MELBOURNE LYING-IN HOSPITAL 1857–1883

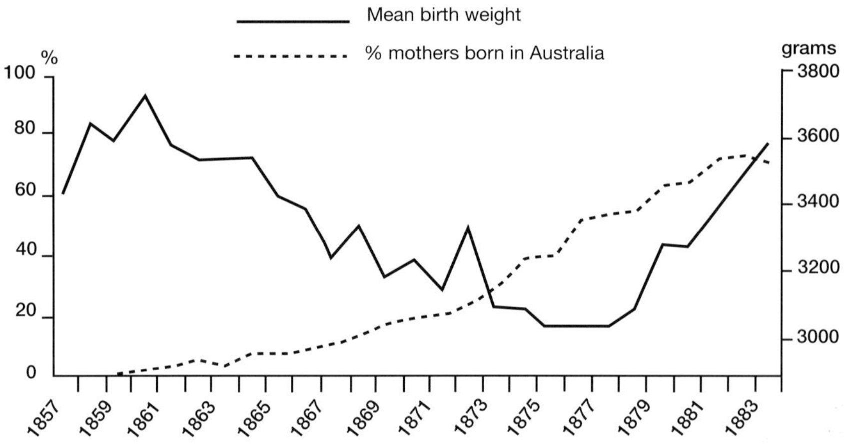

In fact in the other twenty-seven deaths of Lying-in Hospital babies in that fearful summer of 1875, only two were attributed to 'gastro-enteritis'. Most with diarrhoea had suffered it for too long for the cause to be merely infection, although infection could well have been superadded. Two babies died from the sequelae of measles, and three from chest infections. But most continued to die from wasting and ill-nourishment: that is marasmus and inanition, and prolonged diarrhoea. Some were small babies, probably premature; a number possibly 'small for dates' in that they were light but of normal length which may suggest poor growth in utero. But Thomas Albon was another born to term at $7\frac{1}{2}$ pounds [$3\frac{1}{3}$ kg] who died after four months of infantile wasting. (Again his mother was English, and again he was neither illegitimate nor a first child, though his parents disappear from the Victorian record, having either changed their name or moved on.)

We have death certificates for twenty-eight babies who were born in the Lying-in Hospital and died either in January or February 1875. Of the mothers delivering in the hospital during 1874, 40 per cent were colonial born, 60 per cent born overseas. The birth weight of colonial-born mothers' babies was significantly lower than that of the children of overseas-born mothers. Indeed between 1857 and 1883, Scottish mothers had the heaviest babies, Irish women came next, then the English, then the Victorian, and fully half a pound behind the Scots and Irish came the babies of Tasmanian mothers.

The reasons behind what at first sight looks an anomalous finding — that the colonial born reared on plentiful protein and sunshine should be suffering a level of

TABLE 2
MEAN BIRTH WEIGHTS BY MOTHERS' PLACE OF BIRTH,
MELBOURNE LYING-IN HOSPITAL, 1857–1883

Mother's country of birth	Number/(per cent)	Mean birth weight in pounds (grams)
Australia	**2976 (33 per cent)**	**7.1 (3224)**
Victoria	2217 (25 per cent)	7.1 (3240)
Tasmania	354 (4 per cent)	6.9 (3141)
Other colonies	405 (4 per cent)	7.1 (3208)
British Isles	**5823 (65 per cent)**	**7.5 (3388)**
Ireland	2985 (33 per cent)	7.5 (3413)
England	2228 (25 per cent)	7.3 (3339)
Scotland and Wales	610 (7 per cent)	7.6 (3449)
Other	**215 (2 per cent)**	**7.2 (3272)**
Total	**9014**	**7.3 (3331)**

nutrition and health in pregnancy so much lower than that of women who had at some time experienced the dislocations of emigration — are complex and have everything to do with the persistence of a colonial underclass of immigrant failures and convict heritage.[19]

Our collection of twenty-eight infant deaths largely fits the wider analysis of the comparative biological fates of the newborns. First the 'Tasmanians' and the 'migrants' were slightly over-represented. The colonial-born mothers were more likely to be single, as were all the Victorians (seven out of seven) and one of the Tasmanians (out of four), with one doubtful who clearly would not disclose her maiden name in the hospital, and whose putative husband was designated 'gentleman' on the baby's death certificate. Her son started life as Reginald and concluded it as Charles, dying at seven months from 'bad nursing and want of breast milk' leading to 'inanition', or 'emptiness', over fourteen days. In stark terms, he starved to death. His mother was just twenty, far from home, living in Little Collins Street, perhaps now abandoned by her 'gentleman'. She then disappears from the Victorian public record.[20] The Tasmanians and the Victorians were young, all but one being under twenty-two, four of them adolescents. Only one, a Tasmanian, Mary Ann Sherrell (née Roberston), wife of a Collingwood cab owner, was older at twenty-six and Robert was her third child. But she was mortally ill from 'disease of liver and kidneys', possibly the result of alcoholism, and died a pauper's death in the Melbourne Hospital, eleven days before Robert died at nine months, after ailing for five months from 'catarrh or muco enteritis'.[21] Agnes Summers' premature baby died at eighteen days and a coronial inquest found that Alfred died from 'congestion of the brain and membranes' due to 'improper and unsuitable food': in other words he was not breast fed, may have been fed solids and alcoholic stimulants. He was born at 4½ pounds [2 kg], but a long 17 inches [43 cm], which suggests both prematurity and poor growth in utero.[22]

Of the seven Victorians, all but two had babies who died under the age of five months. All the deaths could be explained by poor care from young, distressed, unsupported single mothers. The oldest baby, Caroline Hildebrand, survived until eight months before succumbing to pneumonia. Her 18-year-old single mother, living in the heart of the red light district around Little Lonsdale Street, had done well to keep her alive for so long.[23]

The story is different with the seventeen infants whose mothers were born overseas. First, six of the mothers were married and one married soon after the birth; second, all were over twenty-one years of age and significantly more mature. Indeed three of the single mothers were much older Irish women: Ellen Delahunty was thirty-four and lost both her son Pierce in 1875 and a daughter born the following year. The Irish accounted only for six of the seventeen, although three of the English had Irish names and may have been on a second migration leg after leaving Ireland. Significantly, their babies were born bigger and stronger (eight out of fourteen weighed 7 pounds [3.18 kg] or more, with just one clearly premature and eight possibly small-for-dates). Most important of all, thirteen of them died after their fourth month, succumbing in the summer after surviving the perilous neonatal period. Of the mothers whose subsequent child bearing could be tracked, only one — Ellen Byrnes, wife of an illiterate labourer — reared most of her children successfully, losing only four out of ten (including twins born last, when she was thirty-six).[24] It is these deaths that more significantly suggest that extreme weather may have been the deciding factor that pushed these children over the edge. Their mothers' inexperience with hot weather may have contributed to their collapse.

Improper diet

As mentioned before, few of the cases look like classical gastro-enteritis. The babies had been suffering from diarrhoea or mucous stools for too long for the cause to be only infection or even toxicity from *E. coli* contamination. Instead the diarrhoea and its duration prompted doctors at the time to blame 'unsuitable food' or 'improper feeding'. The baby's digestive system would be too immature to cope and the indigestible food would be expelled with diarrhoea and mucous. The child would fail to thrive and fade away. This might not just be hand feeding cow's milk, dangerous as that was of itself, but also the feeding of solids and stimulants, even to the newborn. Not all instances were so blatant as to initiate a coronial inquiry, but one extreme case in 1864, discussed by Dr Gerald Fetherston of the Lying-in Hospital at a clinical meeting of the Medical Society of Victoria, is illustrative because of the post-mortem findings:

> The child had been born at the full time in the Lying-in Hospital. The mother was an extremely intemperate woman, and as soon as she was discharged from the hospital, commenced drinking excessively. Four days prior to the child's death, she had ceased to suckle it, and a neighbour having undertaken to attend to it, gave it maizena and sago. The child was excessively sick, and appeared to suffer pain; the diet, however, was continued, and a dose of castor oil was administered. Death occurred suddenly, while the child was being placed on the mother's lap, and was stated to have been the immediate result of a convulsive seizure.

The post-mortem revealed a ruptured stomach, but the doctors were more concerned with the long-term effects of both ingestion of alcohol through the mother's milk and the strong evidence from the liver and stomach that spirits — gin or whiskey — had been given direct as a soporific. Alcohol had 'softened the coats of the stomach'; there was an excess of bile as the tiny liver struggled to process the toxins. The conclusion was to condemn not just the practice of sedating infants with alcohol, but also the tendency to 'gorge them without stint, with coarse indigestible matter, instead of simply giving them cow's milk'.[25]

Fetherston's paper in the *Australian Medical Journal* was followed by a long, careful article 'On the Mortality of Infants in Victoria', presented by Dr Charles McCarthy. McCarthy's analysis would be superseded by the sanitarians and the germ theorists of the 1870s and 1880s, and it would tend to be disregarded by modern historians for its emphasis on poor mothering and domestic management. But it deserves a re-examination, especially if we understand it with some of the perspectives that medical anthropologists have developed. McCarthy set out four causes in order of importance. First came 'improper diet as to *quality* or *quantity*', arguing that his analysis of infant mortality cases in Victoria pointed to 'error in diet' as the 'most frequent' cause of death. Second, he identified 'improper clothing ... setting at defiance the climate and seasons'. Third was impure air in places afflicted with poor drainage and noisome wastes; and fourth was insufficient medical attendance.[26]

Historians of motherhood have advised us to see doctors' attacks on women's practices and lay medical knowledge as self-serving professional boundary setting, but the prolonged improper feeding of solids and soporifics — and of course contaminated cow's milk — makes more sense of most of the infant diarrhoea we find reported in the death certificates of the 1860s, 1870s and 1880s. McCarthy fulminated against untrained midwives and grandmothers who gave newborns gruel, butter and even 'half-and-half' of porter and ale a few hours after birth. Another practice was the tight bandaging and swaddling of babies — a disastrous practice in a hot climate, and even occasionally fatal if the bandaging was too tight.[27] Too many of the babies born to term at the Lying-in Hospital in its first forty years died of marasmus or inanition within a few months of birth for one not also to suspect neglect, ignorance, weak maternal bonding and improper feeding. Only some died on the premises of nurses or baby farmers. The majority died with their mothers, or at least their mothers were present to sign the death certificate. And it is possible that those who died and were not registered, were more likely to have been with their mothers or grandmothers than in the care of baby farmers and institutions.

What we know in retrospect is that gastro-enteritis can cause a baby to develop lactose intolerance, so that even after the illness has retreated, the baby cannot absorb enough nutrition from cow's milk to recover condition and begin to thrive again.[28] Acquired lactose intolerance and subsequent wasting remain a problem in Indigenous health in this country.

Overheating

It is Dr McCarthy's second and third causes — excessive clothing and impure air — that bring us back to hot winds and extreme weather. And here health beliefs are also significant. When doctors and health officials deplored the ill-ventilated, suffocating dwellings and workplaces of the people, they revealed how much colonial Australians, in particular those who had grown up in Great Britain or Ireland, still feared 'cold air' and 'getting wet' and shut out the 'night air' as the bringer of 'pestilence'. Dr James Neild wrote eloquently in 1861 of the poor ventilation of colonial buildings, of the many domestic bedrooms that had no windows or ventilation at all other than the door. He correctly perceived the danger of poor ventilation to phthisical disease, but he also noted the misery these new colonials inflicted on themselves in hot weather:

> Our churches and chapels are faulty alike with the rest of our public buildings. There is no contrivance for carrying off the foul air save by means of the windows, which are almost always closed, to allay the apprehensions of those who live in dread of draughts, and on a seething night in January the sad spectacle may often be witnessed of five or six hundred persons suffering with exemplary fortitude the oppression and distress of breathing an atmosphere baleful with accumulated poison.[29]

Neild deplored the poor quality of window technology in colonial houses, how many built quickly in the gold rush period lacked sashes, ropes and pulleys that made windows easy to open and keep open. But his was only one of many voices, medical and official, which deplored the preferences of the poor for closed, unventilated rooms, especially for children. Dr William Crook testified before the Royal Commission of Diphtheria in 1872 as to the:

> close packing of children in a bedroom at night ... [I am] in the habit of making very early visits to my patients, especially those connected with the friendly societies, and I am constantly driven back from the bedrooms on opening the door first, by the horrible stench and odours from the rooms. In some cases the children have been affected with diarrhoea in the night, the evacuation is not removed, windows and doors closed under the notion derived from the old country that draughts and cold are inimical and are bad at night.[30]

Dr Charles Louis McCarthy as Health Officer for Footscray in 1889 noted that local children suffered particularly because the stench of noxious industries, made all the worse in summer, forced local families to keep all windows closed.[31]

What the doctors did not understand was that it was not so much the noxious vapours that were pernicious, as the overheating, loss of oxygen and especially dehydration that came with being over-wrapped and kept in hot rooms. Dr Mackenna, so concerned in 1858 about the dangers of hot winds to the colonial body, had discovered empirically that cooling sick and distressed children often led to a dramatic improvement in their condition.[32] What makes sense to a modern medical practitioner is that prolonged hot weather, insufficient means of cooling and a shortage of fluid could tip a baby, who had suffered from long-term diarrhoea and improper feeding, over the edge. The disturbance to electrolytes from prolonged diarrhoea and insufficient nourishment and fluid could cause an infant to overheat fatally, suffer convulsions, or become further weakened in the face of new infections.

If new colonists let their morbid fear of draughts inhibit the natural desire to cool the body in hot weather, another practical difficulty was the actual shortage of cooling drinks. The Yan Yean water supply could not match the growth of Melbourne by the end of the 1860s, and on very hot days the water supply would fail. In 1872 during a hot spell lasting five days, the water supply stopped altogether for three whole days and nights. Even the main suppliers of alternative drinks — the breweries — had to stop production.[33] In that terrible January of 1875, again the water stopped, leaving 'women and little children ... crying for water while the atmosphere was like the mouth of a furnace'.[34] Indeed one of the hardships of colonial life was the deficiency of palatable non-alcoholic alternatives to water and beer, especially in the summer.[35]

Adaptation to a new climate took time, in particular when adaptation entailed the discarding of ancient health beliefs and nostrums and accepting the advice of professionals and experts, with all the loss of culture and personal autonomy that such cultural changes demanded. This was one instance where the constitutionalists among the medical profession could observe more accurately the benefits of simple changes to care and practices that would enable colonial bodies to acclimatise more efficiently. As Dr James Jamieson declared in 1878: 'It is a slow process that of educating a community'.[36]

Janet McCalman

Janet McCalman is Professor in the Department of History and Philosophy of Science at the University of Melbourne and Director of the Johnstone-Need Unit for the History of Medicine in the Centre for the Study of Health and Society. She was elected a Fellow of the Australian Academy of the Humanities after being awarded its inaugural Max Crawford Medal in 1992. She has written three award-winning books in Australian social history. The first two, *Struggletown* and *Journeyings*, explored the social history of Australian private life via generational slice history. Her most recent book, re-issued in the United States by Johns Hopkins University Press, is *Sex and Suffering: Women's Health and a Women's Hospital 1856–1996*. She edits *Health and History*, the Journal of the Australian Society of the History of Medicine.

10

Sunshine, social factors, solar energy and sustainability

Ian Lowe

Some social context

The image we have of Australia is inescapably linked to the sun. Among my earliest memories of formal education are the times spent reciting the jingoistic jingles of the old poets: 'I love a sunburnt country', 'He sees the vision splendid of the sunlit plains extended' and so on. The sun was also a crucial factor in the youth of my generation, with an enduring legacy. The sun-soaked summers of school years flowed into the sun, sand and self-conscious sexual experiences of early adulthood, leading inexorably to the marriages of mature years and the melanomas of middle age. As the only place in the world where large numbers of inadequately pigmented people expose their bodies to the tropical sun, Australia is inevitably the country with the world's highest rates of skin cancer, increasing from about 20 per cent of adults in southern areas to a remarkable 50 per cent in northern parts of Queensland.[1]

Growing up in a time when politicians spoke openly of their desire for a 'white Australia' and the *Bulletin* could proclaim on its cover the credo of 'Australia for the white man', I was struck by the dedication my fellow white Australians put into becoming as non-white as possible, with the bronzed sun gods of the surf lifesaving clubs seen as the ultimate goal. Even Australia's consistent ability to punch above the weight of our small population on the sporting field has been attributed to a healthy outdoor culture and plenty of sun.[2]

An equally important cultural metaphor was the determination of the predominantly European immigrants to tame this 'bright and savage land', to overcome the forces of nature and turn the Australian bush into an antipodean reconstruction of European managed lands. Victoria even to this day calls itself 'the garden state', a remarkable metaphor for anyone standing in the mallee, the Gippsland forests or the Alpine scrub. The mission of our forebears was to clear the land of native vegetation, dam rivers, build harbours, plant introduced species and make the land 'productive' in a narrow economic sense. Given the technology of the time, such as axes and cross-cut saws, the scale of the change to natural areas was nothing short of heroic. A 1925 film, celebrating the results of years of dedicated effort by the pioneers, concluded its litany of changes to natural areas with the ringing declaration, 'The Earth yields to the dominion of man'.[3] This is the social and cultural background against which we must measure our reluctance to use solar energy; our love of the sun is overpowered by our desire for a managed energy system, rather than one dependent on the whimsical variations in natural energy flows.

Energy and society

The society we now live in is energy-intensive, requiring enormous amounts of fuel energy. The national rate of energy use is roughly equivalent to about 6 kilowatts per person.[4] This very large total reflects the fact that energy is not just used in the ways we notice when we switch on a light or put fuel into a vehicle, but is a vital input to all the goods and services of modern society: buildings, transport vehicles, clothes, sporting equipment, food and water all require energy. Mining and mineral processing to produce our basic raw materials use huge amounts of energy. One way of quantifying our energy use is to equate it to each Australian having the services of about fifty human slaves working in shifts.[5] Energy does for us what slaves once did for despotic rulers; it carries us about, washes our clothes, cooks our food, entertains us, provides us with hot water, fans us when we are hot, disposes of our wastes, and so on. Further, energy is the key to other possible shortages. Energy can be used to farm land more intensively and produce more food from the same area. It can be used to obtain potable water from the sea if we exhaust available fresh water. If sufficient energy is available, we can process poorer grades of ore to obtain the minerals we want as richer ore deposits are exhausted. Without usable energy, our society would literally grind to a halt.[6] So a crucial issue is the sustainability of the present pattern of energy supply and use.

The Council of Australian Governments adopted a national strategy for ecologically sustainable development in 1992. This declaration nominally committed all state and territory governments, as well as the Commonwealth, to developing in a way that would be sustainable.[7] Some operational criteria to determine whether an activity can be sustained are clear. Will it lead to depletion of resources? Will it do environmental damage? Will it cause social stress? Is its economic viability assured? The difficulty is that all of these criteria are subjective. There is no scientific principle that determines whether the level of resource depletion, environmental damage or social stress falls within the range that will be sustainable, while economic analysis is legendary for its dependence on the assumptions used.

Those qualifications aside, it would take a creative advocate to make a convincing case that the present approach in countries like Australia is sustainable. At the global level, the second United Nations Environment Program report on the global environmental outlook concluded that the present approach is not sustainable.[8] Arguably, it fails all the criteria listed above. In that sense, 'doing nothing is no longer an option'.[9] A commitment to sustainability requires a radically different approach to energy.

Resources

There is no absolute shortage of energy resources. The known reserves of most fossil fuels are very large. In orders of magnitude, the known coal is sufficient to keep using it at the present rate for more than 1,000 years. The known reserves of natural gas are large enough to sustain about 100 years use at the current rate. That may be a misleading assurance, as the use is increasing at about 5 per cent per annum, equivalent to a doubling about every fifteen years.[10] If that rate of exponential growth were to continue, there could be resource problems within thirty years. Of more immediate concern is the case of oil, the basis of our entire transport system as well as the feedstock for the petrochemicals industry. The best estimate of the peak in world oil production is about ten years away.[11]

The current expectation that we will be able to continue our profligate use of petroleum fuels for several decades is based on a presumption that the present inequalities in access to transport services will continue, Australian families having several large cars in a world where the majority of people have never even ridden in a car. As United States commercial media encourage the global expectation of mobility, the pressure will grow for a more responsible and equitable use of the remaining petroleum resources. It would be wise to assume that we will need to develop a radically different approach to transport within a relatively short period. We should be reassured that the available renewable resources are very large by comparison with our energy needs. In orders of magnitude, the solar energy that hits Australia in one summer day is similar to the global energy use for a year.[12] In other words, the rate at which we receive solar energy is about 10,000 times our present energy use. Any conceivable future level of energy use is a tiny fraction of the available solar energy alone, without considering such other sources as wind, tidal, geothermal and wave energy.

Environmental impacts

There are several local environmental problems associated with the extraction, processing and use of fossil fuels, most obviously the resulting pollution of air and water. The most serious problem is its contribution to global climate change, identified by the United Nations Environment Program as one of the five most serious environmental issues facing human society.[13] The trace gases in the atmosphere are responsible for a natural 'greenhouse effect', making the Earth about 33 degrees warmer on average than the Moon.[14]

The Industrial Revolution began a process of using fossil fuels (coal, oil and gas) to obtain our energy, releasing the stored carbon into the atmosphere in the form of carbon dioxide. The amount of carbon dioxide in the air is now about 30 per cent greater than it was 200 years ago.[15] While the geological record shows that the level has varied between 180 and 280 parts per million during the last half million years, the 2003 level was 380 parts per million. Scientific opinion now strongly supports the view that the global climate has changed significantly as a result, with the average temperature about 0.7 degrees higher than 100 years ago and related changes to rainfall patterns, sea level, areas of snow and ice cover, growing seasons, and so forth.[16] Of a range of possible future scenarios developed by the Inter-governmental Panel on Climate Change (IPCC) in its Third Assessment Report, the most optimistic is that the temperature will only increase by a further 1.5 degrees in the next century, with pessimistic outlooks giving temperature increases two to four times as large.[17] More fundamentally, the IPCC has also warned that even these quantitative assessments may be optimistic:

> Future climate change may involve 'surprises'. In particular, these arise from the non-linear nature of the climate system. When rapidly forced, non-linear systems are especially subject to unexpected behaviour.[18]

This assessment has produced a growing awareness of the need to curb emissions of carbon dioxide and other greenhouse gases, leading to the negotiation of the Kyoto Protocol.[19] This agreement commits the industrialised world as a whole to reducing its emissions of greenhouse gases to 5.2 per cent below the 1990 level by the 2008–2012 period. Within that global goal, different countries have different targets. Japan agreed to reduce emissions by 6 per cent, Canada and the USA by 7 per cent, while the European Union agreed to cut its total emissions by 8 per cent. Within that overall target for Europe, such countries as Germany and the UK will make greater savings to allow greater levels of energy use in more agrarian nations like Portugal and Greece. Four countries were recognised as special cases and given more generous targets in the Kyoto Protocol. Norway, New Zealand and Iceland were given generous targets because they already get more than 85 per cent of their electricity from renewable sources — wind, geothermal and hydro-electricity — and so have limited opportunity to make savings. Australia was given a generous target of increasing emissions by 8 per cent above the 1990 level because the government delegation insisted it would only join the agreement if given this special treatment.[20] Further, the Kyoto meeting agreed in the early hours of the last morning to add what is known around the world as 'the Australia clause', a provision that both the 1990 baseline and the 2008–2012 target would include changes to land use as well as the burning of fossil fuels. The Australian delegation's argument was cynical in the extreme. Large-scale land clearing in Australia was emitting huge amounts of carbon dioxide in 1990, about one-third of all emissions. Including land clearing in the baseline means, therefore, that even with an increase in energy-related emissions of up to about 40 per cent, Australia can still meet its Kyoto target provided land clearing is halted by 2008.[21] Recent statements by government ministers declaring Australia 'on track' to meet the Kyoto commitment, despite continuing rapid growth in energy-related emissions, are based on this creative arithmetic.[22]

Kyoto was seen at the time as a landmark agreement, with all industrialised nations agreeing to set some limits on emission of greenhouse gases. Since then, the installation of George W. Bush in the USA has seen his regime renege on that country's commitment to Kyoto, with the Howard government in Australia predictably following the same line. With growing awareness of the reality of climate change, there is likely to be increasing pressure to move beyond the Kyoto agreement to serious emission reductions. The science shows that global emissions will need to be reduced to about 40 per cent of the present level just to stabilise the global atmosphere at a new level, higher than at any time in the existence of the human species. If this is to be achieved in ways that allow poorer nations to achieve legitimate aspirations for the basic features of a civilised lifestyle, the affluent countries will have to make serious reductions in their emissions. There is likely to be severe pressure on Australia, as the nation with the highest per capita emissions in the industrialised world, as well as a range of policies that still actively promote wasteful use of energy. A study released by the Australian government in 1996 estimated the annual financial subsidy of fossil fuel supply and use at $2 billion, with uncharged environmental costs of $4 to $5 billion in addition, without attempting to quantify the cost of climate change.[23]

There are three ways to reduce emissions. We can reduce energy use, improve the efficiency of converting energy into the services people want, and use less carbon-intensive sources of energy.

There is relatively little scope for using less energy because fuel energy provides the quality of life that most people now take for granted. We will probably move in future to eliminate wasteful uses of energy, such as car races or lifts to go down, but our society is likely to remain energy intensive for the foreseeable future.

Much can be done to improve the efficiency of turning energy into the services people want: hot showers and cold drinks, comfort and mobility, the capacity to cook food and wash clothes. Australia has finally introduced minimum energy performance standards for some domestic appliances, such as refrigerators and electric hot water systems. While these standards are a long way from international best practice, they will still produce a significant improvement in energy efficiency.[24] We can assume that almost all appliances in use by 2012 will meet the new minimum standards. This will significantly reduce emissions from such appliances as refrigerators, as well as electric hot water systems if such wasteful technology continues in use. Australia recently hosted an OECD Eco-Efficiency conference at which a wide range of case studies showed that reducing resource use to a quarter of the present level is possible using technology that is cost-effective today.[25] Only the most naive economists doubt the scale of savings that can be achieved at a profit. We should be moving to toughen up the minimum standards to reflect world's best practice. We should also be extending the minimum standards to a wider range of domestic appliances, and including basic industrial equipment.

In 2000, Robert Watson, chair of the IPCC, told the parties to the UN Framework Convention on Climate Change that cost-effective reductions in greenhouse gas emissions can be achieved in the areas of energy supply, energy use, waste management, agriculture and forestry using existing technology.[26]

> The optimum mix of policies to facilitate the penetration of lower carbon intensive technologies and encourage the efficient use of energy will vary from country to country as policies need to be tailored for local situations and developed through consultation with stakeholders. Policies include energy pricing strategies (e.g. carbon taxes and reduced energy subsidies), reducing or removing subsidies that increase greenhouse gas emissions (e.g. agricultural and transport subsidies), internalisation of environmental externalities (e.g. incorporating the health costs associated with particulates caused by combustion of coal into the price of coal), incentives such as accelerated depreciation and reduced costs for the consumer, domestic and international tradable emissions permits and joint implementation, voluntary programs and negotiated agreements with industry, utility demand-side

> management programs, regulatory programs including minimum energy efficiency standards, market pull and demonstration programs that stimulate the development and application of advanced technologies, and product labelling.

Watson went on to argue that policies which address simultaneously local or regional environmental problems, such as air quality or land degradation, are particularly attractive. 'Significant reductions in greenhouse gas emissions can be accomplished by pursuing sustainable development goals', he concluded.[27]

Cleaner energy sources

The Australian government set one concrete target in response to global warming when it proposed to require electricity retailers to obtain by 2010 an extra 2 per cent of their supply from renewable sources.[28] That was interpreted to mean we could expect 2 per cent of the total coal-fired electricity used in 2010 to be replaced by renewable energy technologies. This timid response was diluted further in response to arguments of sectional interests, mainly coal-fired electricity generators and subsidised large-scale power users. At the time of writing, a review of electricity market arrangements by a hand-picked group of two economists and one coal executive had actually recommended that even this inadequate target should be scrapped in favour of 'market mechanisms'.[29] Since that approach fits comfortably with the ideology of the Howard government, it seems likely to be adopted.

A serious target would be to get 10 or 20 per cent more electricity from renewable energy technologies as part of the process of moving to energy forms that emit less carbon per unit of delivered energy. In round figures, oil emits about 80 per cent the carbon per unit energy of coal, gas about 60 per cent, coal-fired electricity about 300 per cent and renewables like solar energy about 10 per cent.[30] So replacing coal-fired electricity with almost any other energy supply makes sense in terms of reducing emissions. Cooking with gas rather than coal-fired electricity emits a small fraction of the carbon dioxide to achieve the same task. Co-generation schemes, or combined heat and power schemes as they are known in Europe, roughly halve the emissions per unit of useful energy, so we should be developing measures to encourage that technology, such as the schemes to use 'waste' heat from sugar mills.

The biggest gains occur when coal-fired power is replaced by renewable forms such as solar, wind or biomass. Many households and some businesses have now adopted green power schemes, paying a premium to have electricity entirely from renewable sources. This is a remarkably cost-effective way of achieving large reductions in emissions. It adds about 20 per cent to the power bill to get all of my electricity from renewables. I calculate that a typical Brisbane household using the local 'Earth's choice' scheme reduces its carbon emissions by about as much as giving up the family car, for a cost of about $2.50 per week. Such schemes should obviously be promoted. A larger fraction of overall electricity from renewable sources is an option that should be considered. Extrapolating from the costs of existing green power schemes, the government's target of 2 per cent would add about 0.4 per cent, or a few cents a week for a typical household. That is so small as to be undetectable. A more serious target, say 20 per cent extra renewable power, would only add 4 per cent if the present experience could be scaled up; since we have so far used the most economically attractive power schemes a more realistic figure would be two to three times that, or about 10 per cent. This sort of increase is clearly considered politically acceptable, because it is the increase the government chose to apply to most goods and services with its Goods and Services Tax (GST). If such a policy were to be implemented, governments would probably find it politically expedient to continue to provide public subsidies to such energy-intensive activities as aluminium smelting which would not otherwise be economic.[31] The imposition on most areas of business would be less than the GST, as the reason most businesses pay little attention to energy efficiency is that energy costs are a small fraction of a typical operating budget.[32]

Many overseas countries have invested heavily in new forms of renewable energy; for example, about a quarter of the installed capacity of the state of California is now made up of various forms of renewable electricity supply, and most European countries have more serious targets like getting by 2010 an extra 10 per cent of their electricity from renewables.[33] The Stanwell Corporation in Queensland had stated publicly in 2001 that it would not build any more coal-fired power stations, with all its envisaged new capacity in various forms of renewable electricity such as wind, solar thermal and waste utilisation, but a recent change of management has produced a more conventional approach to supply.

Energy choices are clearly influenced at the margin by price. Domestic electricity users receive no price information at all at the point of use, as a result of which most have very little idea of the best ways to save energy. So it is likely significant reductions could be achieved by provision of price information. The taxing of energy should also be considered, as the National Strategy for Economically Sustainable Development (ESD) Working Group on Energy Use recommended; there is no rational basis for a system which taxes petroleum fuels but does not tax gas, coal or electricity.[34] Given the huge difference in greenhouse impact if, for example, a domestic or commercial user chooses gas rather than coal-fired electricity for cooking or heating, a government trying seriously to meet an emissions target would be trying to influence those choices. Many countries now have carbon taxes or energy taxes to influence consumer choice. Those options should at least be evaluated. As with any other tax change (including the introduction of the GST), there would be winners and losers, so careful consideration should be given to equity issues.

Solar energy

As discussed above, the fundamental change that is necessary is to move away from fossil fuels to use the abundant direct and indirect forms of energy from the sun. Solar energy can be used directly to heat water for domestic or commercial use. The technology is now mature and used by about 5 per cent of Australian homes. There are huge regional variations in the use of solar hot water, ranging from about 45 per cent of households in the Northern Territory to 25 per cent in Western Australia, 7 per cent in the Australian Capital Territory, 6 per cent in Queensland and about 1 per cent in Victoria. While some of these differences arise from variations in the economic attractiveness of the technology between sunnier and cloudier regions, this is only part of the explanation, as the economic returns are better in Queensland than in Western Australia.[35] Overseas studies show that the acceptability of new technologies like solar energy is largely driven by social and political factors.[36] Since the payback time for the investment in solar hot water is shorter than the guarantee period of new models for most of mainland Australia, it makes economic sense to require new houses to use the technology, as some local authorities are now doing.

Solar energy can be used directly to grow bio-fuels. Firewood is still an important heating source in many parts of regional Australia, while bagasse — the dried residue from crushing sugar cane — is a significant source of energy in sugar-growing regions.[37] Solar energy can be concentrated to provide process heat or to drive turbines for generating electricity. Among a range of schemes being explored in Australia, the most ambitious is a proposal to build a kilometre-high tower to generate 200 megawatts of electricity, using the sun to heat air in a huge greenhouse.[38] Solar energy can be converted directly into electric power by photovoltaic cells, which are mainly used for consumers remote from normal distribution systems. The USA, Japan and Europe have ambitious schemes to install millions of these systems on urban rooftops, from where they would feed power into the grid system during the day.[39] It has been argued that the resulting economies of scale will bring the cost of solar cells down from the present levels of $1 to $4 per watt to about 40c in the next decade.[40]

Some technical issues

As discussed earlier, the scale of the solar energy resource is enormous by comparison with any conceivable human demand for energy. One obvious problem is the irregular rate of delivery. There are cloudy days, and it is certain that solar energy will not be available after sunset. While small numbers of people have solar cells on the roof, they can supply power to the grid during the day and extract power at night, using the grid as if it were a storage system. That approach will not work if scaled up to a significant fraction of roofs. Solar energy can be effectively stored if it is tapped as the energy of wind or flowing rivers, or the energy stored in biomass. These are much more attractive large-scale options than banks of storage batteries. So a varied portfolio of renewable energy technologies, feeding into an inter-connected grid system, offers the best prospect of using clean energy on a large scale. Australia's eastern states now have a connected power grid, making it possible for solar energy from Cape York to be used in Melbourne, or wind energy on the South Australian coast to power after-dark activity in Cairns.

Some social issues

The acceptability of new technology is always a matter of social choice. Studies have shown that the best indicator of whether a household will use solar hot water is the number of people known to the decision maker who already have solar hot water systems installed.[41] As discussed earlier, the large differences between Australian states in use of solar hot water certainly cannot be explained by economics, but there are explanations based on the differing social and political climates.[42] There has been clear hostility to renewable energy technologies among traditional electricity professionals, almost certainly resulting from the desire to manage and control power supply, rather than being forced to adapt to the whim of natural energy flows, as suggested in the introduction to this chapter.[43] It involves moving away from the traditional approach of a small number of large power stations located on coal fields and managed by a technical elite, distributing power to a distant network of passive consumers. The new model will be a system of 'distributed generation', in which large numbers of small electricity production units are embedded in the network of consumers, with a consequent much higher level of local control (and much lower level of control by the technical elite who now run power stations). It is not just a facile pun to describe this as a power issue.

Some political issues

Successive Australian governments have been unenthusiastic about international attempts to control carbon dioxide emissions, with the most credible explanation the wish to protect the coal industry from the consequences of serious cuts.[44] This raises a general political issue. Old industries will always have more political influence than emerging ones, so the political system will always tend to protect those vested interests, as Kennedy argues:

> Since most politicians ... have risen to the top through a process of compromise, making deals ands alliances, and taking care not to annoy powerful interests, they are hardly prepared to endorse controversial policies now for purported benefits twenty years away — especially when experts argue that there is little or no cause for alarm ... or that further study is needed.[45]

When the introduction of a new tax system increased the price of solar energy systems, the Australian Democrats used their controlling votes in the Senate to persuade the government to introduce rebates to make the changes price neutral. The government only provided for the rebates to be available for a limited time, so they lapsed at the end of the 2002–2003 financial year, providing a price incentive to use diesel turbines rather than solar energy.

A political storm blew up late in 2002 when the government decided not to provide funding for a proposed new Cooperative Research Centre for renewable energy, while resourcing three new coal-based centres at a total cost of more than $60 million.[46] Thus it continues to be true that the lion's share of Australian energy research funds goes to the two energy technologies least likely to be of importance in the decades to come, nuclear and coal, while almost derisory amounts of public funding support research into the renewable energy sources of the future and more efficient use.[47] Finally, the moves to put electricity supply into private hands inevitably lead to a narrow emphasis on short-term economic considerations, to the exclusion of the social and environmental issues which promote use of solar energy.

Conclusion

Elaborating on the United Nations Environment Program conclusion discussed earlier, that the present approach is not sustainable, the Global Scenarios Group explored the possibility of a transition to a sustainable future, concluding:

> The analysis suggests that the momentum toward an unsustainable future can be reversed, but only with great difficulty. *The Great Transition* assumes fundamental shifts in desired lifestyles, values and technology. Yet, even under these assumptions, it takes many decades to realign human activity with a healthy environment, make poverty obsolete and ameliorate the deep fissures which divide people.[48]

While human society is on the planetary equivalent of the *Titanic* and steaming toward an ecological iceberg, there are still some urging us to speed up, as if eager to accelerate toward catastrophe. This is a consequence, and a logical outcome, of belief in the pig-headed model of the universe [Figure 1], seeing the economy as the main game, while society and our natural environment are smaller domains of lesser importance.[49]

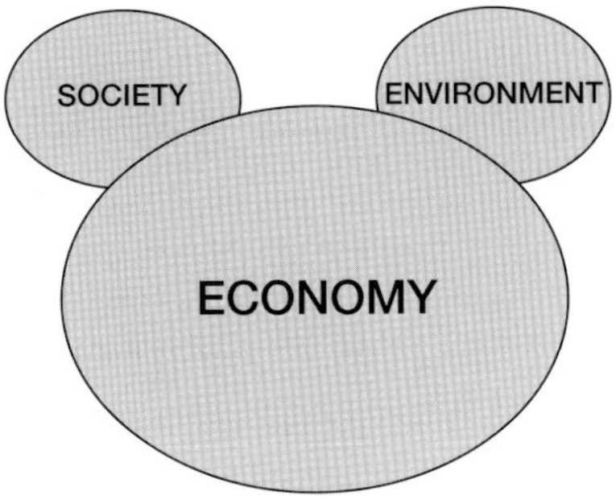

Figure 1
The pig-headed model: the economy is seen as pre-eminent, with society and environment much less important

Those whose thinking is frozen in this out-dated model actually believe that any social or environmental problems can be solved as long as economic growth is sufficiently strong. A recent study of the 1990s by the Australian Bureau of Statistics concluded that all economic indicators had been positive for that decade, but social indicators had been mixed with some very worrying negative trends, while all but one of the environmental indicators had deteriorated.[50] The only model of the world which corresponds to reality is the view-from-space model [Figure 2], which sees societies contained within natural ecological systems on which we depend for breathable air, potable water, nutritious food, cultural identity and spiritual sustenance, while the economy is a sub-set of society.[51] This model leads inevitably to the conclusion that our social and economic planning needs to recognise and work within the limits of ecological systems.[52]

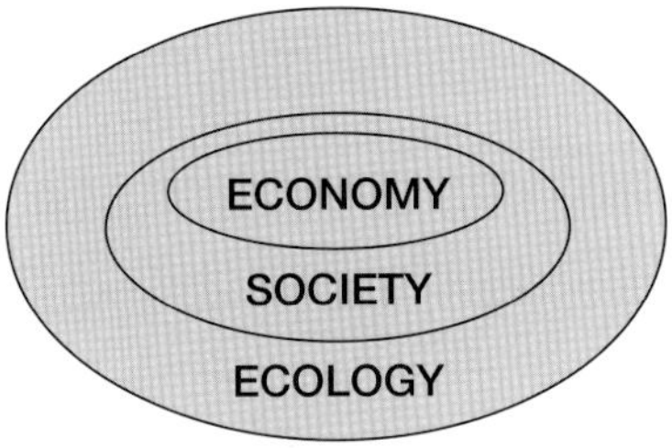

Figure 2
The view-from-space model reflects the real world: the economy is part of society, which exists entirely within natural ecological systems

The sustainable future, powered by renewable energy and characterised by equitable treatment of all humans and a respect for natural systems, remains a Utopian vision. I like to remind people that a world without slavery was a Utopian vision 200 years ago, with those arguing the case being told that no economy could function without slave labour. It was Utopian 100 years ago to be arguing for universal adult franchise. Closer to our time, it was still Utopian twenty years ago to dream of Berlin without The Wall, or South Africa without apartheid, or a treaty to protect Antarctica from mineral exploration and mining, or even good coffee and civilised licensing laws in Brisbane! In fact, most of the social and technical gains that we now take for granted were once Utopian visions, turned into reality by visionaries determined to produce a better world. The inevitable future move to solar energy will represent the victory of our love of the sun over the primitive masculine obsession with technocratic control.

Ian Lowe

Ian Lowe is Emeritus Professor of Science, Technology and Society at Griffith University. He has worked on social and political aspects of energy supply and use for over thirty years. Among a range of advisory roles he directed the Commission for the Future in 1988 and chaired the council that produced in 1996 the first national report on the state of the environment. He writes a weekly column for *New Scientist*.

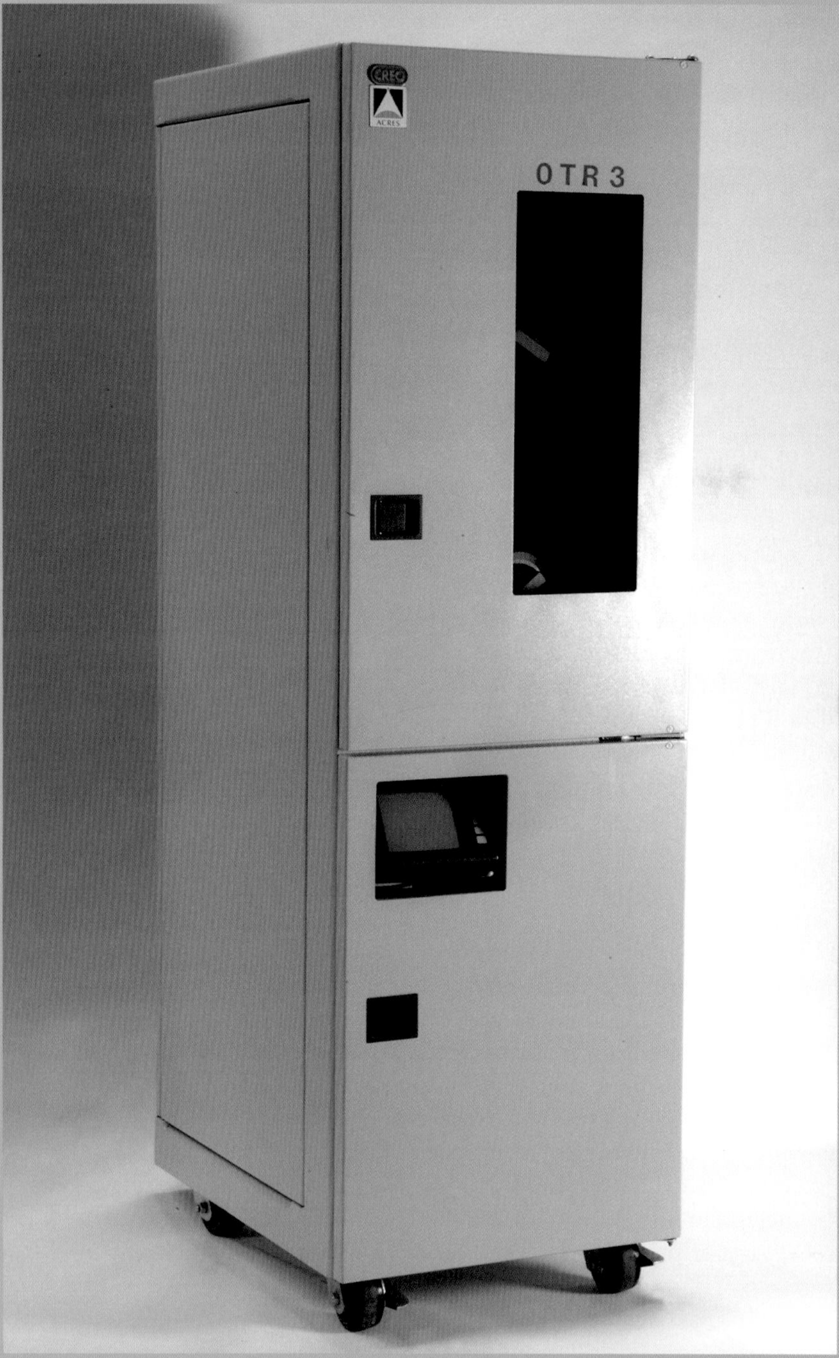

CREO Model 1003 Optical Tape Recorder, about 1990

NATIONAL MUSEUM OF AUSTRALIA
PHOTOGRAPH BY DRAGI MARKOVIC

The Long Range Outlook

A significant challenge for the Australian Centre for Remote Sensing in Canberra has been to find an efficient storage and retrieval system for the vast amounts of remotely sensed data received from a variety of earth satellites since 1979.

The Australian Centre for Remote Sensing is Australia's major satellite remote sensing organisation, collecting and storing data from a number of earth satellites on a daily basis. This data is used in a wide variety of mapping and resource analysis exercises including topographic map making, analysis of vegetation types, natural disaster assessment and the modelling of global climate systems.

Initially a Thorn Magnetic Tape Recorder was used, but in the late-1980s, it was discovered that magnetic tape as a storage medium for digital and analogue information had a shelf life of only about six years. It was replaced by three CREO Optical Tape Recorders in 1990 that remained in operation until 1998. Adoption of the CREO technology brought significant savings in storage space as one optical tape was equivalent to 200 magnetic tapes in capacity. The shelf life of the optical tapes was far superior to the six years for the magnetic tape. The optical tape recorders were replaced by an automatic robotic library system that used Digital Linear Tapes.

11

Revolutionary weather

the climatic and economic crisis of 1788–1795 and the discovery of El Niño

Richard Grove

Australia has above all other places a claim on the epithet, 'the El Niño continent'. It is regularly beset by the El Niño current and Southern Oscillation (ENSO) phenomenon, which links climate anomalies across the globe. During an ENSO event, a spatial pattern of climate fluctuations develops: heavy rains fall on the Pacific coast of South America, whilst the lands to the west of the Pacific Ocean — from Australia and Indonesia through to southern and north-east Africa, and including South Asia — suffer severe droughts. The distinctive thing about the phenomenon is that fluctuations appear in many locations almost simultaneously. The climatologists call the relations between fluctuations 'teleconnections'.[1]

The remarkable correlations and connections between the strength of the El Niño current and Southern Oscillation have been a particular focus in climatology and oceanography since about 1982, but research into historical records can extend data backwards to approximately 1500 AD, with some reliability. From this long data series garnered from the scrutiny of the archival record, some ENSO events stand out as particularly severe. None is worse that the 1788–1795 crisis, which coincided with the first European settlement of New South Wales in Port Jackson, right at the heart of what we now know to be the part of Australia most affected by ENSO. The main perennial drinking water supply, the Tank Stream, dried up. The same event produced very prolonged droughts in South Asia. The severity of the situation warranted comment in contemporary sources in these and other places. Thus some of the earliest documentation of this global climate phenomenon was achieved.

Despite the regularity of claims that the 'latest' El Niño (whichever it may be) 'is the worst in history', by, for example, United States Vice-President Al Gore in 1997–1998 (and the concomitant fears that such events may be linked to global climate change), the historical as well as the prehistoric records indicate that worse events have occurred in earlier eras.[2] Documentary evidence suggests that very much stronger and longer El Niño events have been experienced globally, even in the last thousand years, than those experienced since 1982. The strongest, this chapter suggests, took place during the Little Ice Age between about 1250 AD and 1900, and especially in the closing years of the eighteenth century, when an El Niño event of exceptional strength and length appears to have affected the entire global environment. This last event is of particular interest in understanding the economic and social history of its time. It also suggests the intrinsic value of reconstructing the characteristics and impact of pre-instrumental El Niño events.

El Niño, Australia and the rest of the world

The extent to which Australia's history has been shaped by El Niño events has been under-explored. This chapter considers the severe environmental events that shaped the world at the time of the first seven years of European settlement in New South Wales. Conditions were remarkably contrary to supporting the imperial agricultural vision for the newly annexed land.

On 5 November 1791 Governor Arthur Phillip reported that the normally perennial Tank Stream flowing into Sydney Harbour had been dry for 'some months'.[3] It did not flow again until 1794. The drought had begun, Phillip records, in July 1790 and almost no rain had fallen by August 1791. In a letter to W. W. Grenville on 4 March 1791, he noted that:

> from June (1790) until the present time so little rain has fallen that most of the runs of water in the different parts of the harbour have been dried up for several months, and the run which still supplies this settlement is greatly reduced, but still sufficient for all culinary purposes ... I do not think it probable that so dry a season often occurs. Our crops of corn have suffered greatly from the dry weather.[4]

The settlement of New South Wales was a struggle because of environmental conditions, yet the knowledge base and networks supporting the agricultural venture were growing. As early as 1791, Phillip was chancing a comparison: so dry a season was rare. There are many problems involved in reconstructing the conditions and severity of El Niño events that took place before about 1870, when the period of modern instrumental observation began. But Phillip's hunch (perhaps informed by Aboriginal insights) that this drought was exceptional can, in fact, be confirmed with hindsight.

Corroborating evidence for the severity of the 1788–1795 El Niño event came from contemporaneous reports of drought in southern Africa, South Asia, the Atlantic, the Caribbean and Mexico, with severe famines being reported in some regions. From the early eighteenth century, and even more so after about 1780, the richness of the available historical record, especially in the South Asian tropics, allows us to reconstruct the global footprint of a major El Niño event with very reasonable accuracy and resolution. This is due partly to the growing contemporary detail of Indian weather and population data gathered by the British East India Company. Useful contemporary global weather data from voyages and new settlements at the time are also available, particularly from the Southern Hemisphere, not least from parts of Australia and Oceania from which weather data prior to the 1780s is largely non-existent.

The drought years of 1788–1794 in India were first recognised as having had a global impact by Alexander Beatson, Governor of St Helena. Beatson suggested in

1816 that drought events of 1791 that had occurred simultaneously in different parts of the world (he referred particularly to India, St Helena and Montserrat) had been part of the same connected phenomenon.[5] Beatson had witnessed drought-stricken parts of South India before arriving in St Helena and had also experienced the effects of the 1791 drought in Mysore.

India and the droughts of 1788–1795

The El Niño episode of 1788–1795 was actually the culmination of a succession of unusual weather episodes that had begun in about 1780, and were characterised by extreme events in both temperate and tropical latitudes in Europe and Asia.[6] These years were especially serious throughout South Asia. One year, 1783, which brought famine to almost all India, was memorialised in popular culture throughout India under the name of the 'chalisa'. The word itself, which emphatically associates the Hindi number 'forty' with a particular variety of famine, may suggest a characteristic return interval of forty to fifty years for severe droughts, an interval which is, very roughly, borne out in reality during the Little Ice Age. The social disruption caused by this particular event was a long-term one, since nearly 4 per cent of all villages in the Tanjore district of the Madras Presidency were entirely depopulated in the early 1780s and over 17 per cent in the Sirkali region.[7] Up to eleven million people may have died in South Asia as a direct result of the passage of this event.

The earliest indications of the event are contained in the manuscript records of meteorological observations made for the East India Company by William Roxburgh, a Company surgeon, at Samulcottah in the northern Circars of the Madras Presidency (modern day Samalkot).[8] Roxburgh had accumulated a fourteen-year set of temperature and pressure data from the early 1770s and was thus able to recognise the exceptional nature of the droughts that began about 1789.[9] These droughts had previously been approached in intensity, he reported to the Company, only by those of a century earlier, in 1685–1687.[10] This earlier event is also now believed to have been characterised a year later in 1687–1688, by 'very severe' El Niño conditions in the eastern Pacific. Roxburgh's rainfall figures record the consecutive failure of the South Asian monsoon between 1789 and 1792, with the most severe failure being experienced in 1790 (see Table 1).

TABLE 1
MONTHLY RAINFALL AT SAMULCOTTAH, ANDHRA PRADESH, INDIA,
MAY–NOVEMBER 1788–1792
(IN INCHES AND TWELFTHS OF AN INCH), AS MEASURED BY ROXBURGH

	1788	1789	1790	1791	1792
May	15.4	1	-	4	3.6
June	7.2	6	1.8	4.1	5
July	22.3	6.10	4.9	5.6	6.4
August	12.2	21.1	3.8	-	1.8
September	8.9	1.4	4.8	3.9	7.5
October	5.9	10.1	1.5	3.3	13.11
November	6	1.3	1.2	6.4	-
TOTAL	**77.5**	**47.7**	**17.4**	**26.11**	**37.10**

Of particular note is the indication that the first major rainfall deficit associated with the event was experienced in 1789 in southern India, more than a year *before* similar deficits were experienced, towards the end of 1790, in Australia, Mexico, the Atlantic islands and southern Africa. There is a possibility that the Indian monsoon is an active rather than a passive feature of tropical circulation and that monsoon failure may be efficient in *foreshadowing* El Niño (rather than being predicted by it). It should be noted that some El Niño events, such as that of 1997, appear to articulate only with a failure of the South-East Asian monsoon rather than with a South Asian failure. In the case of the 1789–1793 El Niño event (and the 1685–1688 El Niño event), a failure of the monsoon in both regions appears to have occurred. By November 1792, over 600,000 deaths were being attributed directly to the prolonged droughts in the 167 districts of the northern Circars of the Madras Presidency alone; up to half the population there died in 1792.[11]

TABLE 2
DEATHS FROM FAMINE IN THE MADRAS PRESIDENCY IN 1792

District	Number
Muglatore	141,682
Havelly 1	53,956
Havelly 2	4,874
Peddapore	184,923
Pittapore	82,937
Nandeganah	11,376
Sullapelly	9,018
Poolavam	16,204
Goulatah	12,639
Cotapilly	4,851
Corcoudah	9,035
Ramachandrapuram	7,430
Cottah	7,800
Somapah Villages	2,306
Noozeed	96,210
Char Mahar	16,245

The long drought periods were interspersed by very short periods of intense and highly destructive rainfall. In three days at Madras in late October 1791, 25.5 inches of rain fell, 'more than ... has been known within the memory of man'.[12] Throughout India, the famines of 1788–1794 resulted in very high mortality. In limited areas, such as the Northern Circars, East India Company officials attempted to estimate total mortality statistics. In other regions a much rougher but still useful guide is provided by the figures for deserted village sites. In the Gorakhpur district of Bihar, for example, the 19,600 villages extant in 1760 had fallen to 6700 by 1801, and only a third of the district remained under cultivation. Not all of these desertions were due to famine mortality, but a high proportion of them probably was.[13]

A similar pattern of mortality obtained in southern India, where a pronounced pattern of village desertion can be established, and up to 30 per cent of villages were deserted, for example, in some parts of Salem district.[14] Extrapolating from these kinds of figures we may attribute a total famine mortality during 1788–1794 of perhaps eleven million to the extended El Niño conditions of the period. However, although the human cost of the episode was particularly high in the sub-continent, severe consequences were also felt elsewhere, especially further east.

El Niño patterns east of Australia to the Caribbean

The rainfall deficiency of South Asia spread towards the east, with unseasonably severe droughts being experienced in Java.[15] In the Pacific region, temperatures were unusually low. There is some limited evidence of cold and severe drought in the western Pacific from the journals of D'Entrecasteaux, who visited New Caledonia in 1793.[16] Another source of data from the late 1780s is the logbook of HMS *Bounty*. Temperature readings were made every four hours while at sea, and the lowest and highest temperatures were read while in port. Bligh wrote while his ship sheltered at Matavai Bay, Tahiti, that:

> I experienced a scene of today of Wind and Weather which I never supposed could have been met with in this place. The wind varied from ESE to the NW and the Therm. stood between 78 and 81.5 degrees. By sunset a very high breaking sea ran across Dolphin Bank, and before seven o'clock it made such a way into the Bay that we rode with much difficulty and hazard. Towards midnight it increased still more and we rode until eight in the morning in the midst of a heavy broken sea which frequently came over us. The Wind at times dying away was a great evil to us for the Ship from the tremendous Sea that broke over the reefs to the eastward of Point Venus, producing such an outset thwarting us against the surge ...[17]

Bligh's observations of atypical weather conditions for the time of year suggest that by December 1788, the El Niño episode was already well under way. A few months later the El Niño might have actually saved the life of Bligh and the men who were cast adrift by the mutineers. Instead of the very hot conditions that would normally have confronted the 23-foot open boat between Tofua in the South Pacific and Timor between 29 April and 14 June, Bligh and his men encountered cold conditions throughout the voyage. Furthermore instead of a dry heat that would probably have been deadly, the rainfall they experienced was so cold that Bligh instructed his men to soak their clothes in warm seawater and then wear the wet clothes to keep warm! The next month, Bligh again recorded heavy rain, 'which enables us to keep our stock of water up'.[18] The crew all complained of rheumatic pains and cold. Furthermore, the supply of rainwater allowed Bligh to avoid making landings on hostile Pacific islands for water, as they would have in a normal year. So the El Niño that began in 1788 probably meant that the history of Captain Bligh and the Mutiny returned to Britain, instead of becoming an uncertain note about the 'mysterious disappearance of both the *Bounty* and Captain Bligh'.

In Mexico the prolonged aridity that developed during 1791 was recorded in the steady fall in the level of Lake Patzcuaro between 1791 and 1793, giving rise to disputes over the ownership of the new land that unexpectedly emerged.[19] As in Europe these events were preceded by summer crop failures. On 27 August 1785 a hard night frost and the ensuing crop failure precipitated the great famine of 1785–1786.[20] In the 1790s in Mexico not one annual maize crop yielded an abundant harvest. This was entirely due to droughts every year, primarily in June and July. The severest droughts of the 1788–1795 El Niño event did not strike Mexico until 1793, so that the onset of full El Niño conditions did not affect rainfall for more than two years after the same event had caused monsoon failure in India. But wholesale failures of the wheat and maize crops took place in 1793 and 1797.[21] In 1794 the maize crop was very poor, due to almost complete drought. In 1795 the crop returned to near normal, although one might note that drought conditions persisted in that year in many Caribbean islands.[22]

Along the Peruvian coast the great strength of the El Niño current itself during 1791 with its resulting degradational effects on agriculture and fisheries was also documented by contemporary observers. Flooding of normally arid areas was especially widespread during that year.

The first indications of the onset of unusual drought in the Caribbean were felt in the most drought-prone islands of the Eastern Antilles, especially on Antigua and Barbuda. Antigua had already suffered from a long drought in 1779 and 1780 in an earlier El Niño episode. In 1789 the drought occurred again, but with 'redoubled severity'. Even as late as 1837 this year was still referred to by Antiguans as 'the year of the drought'. As a chronicler notes, 'What miseries the Antiguans then suffered I am of course from experience unable to say; but if they exceeded those endured in that eventful year 1837 [a later severe El Niño] they must have been terrible indeed'.[23]

By August 1791 the desiccating effects on the islands of the Antilles were already the severest recorded since the late seventeenth century. On the islands of St Vincent and Montserrat no measurable rainfall had been recorded by the middle of the month, according to the colonial archival documentation. Landowners made formal requests for tax relief due to harvest failure.[24] The drought continued on Montserrat until at least November 1792.[25] The timing varied from place to place. For example, the extended and abnormal periods of drought (1791–1794) on St Helena in the South Atlantic were later than those in the Caribbean. On St Vincent and on St Helena the droughts drove calls by government naturalists for the formal gazetting of forest reservations to encourage rainfall.[26] The great El Niño event continued later in the eastern Caribbean and Atlantic, with the *Times* index indicating that in 1795 there was still an unrelieved drought in Antigua.[27]

African droughts and El Niño

The recorded incidence of the El Niño current, as documented by Quinn and Neal, may not be, by itself, an adequate guide to the true dimensions, impact and longevity of the 1788–1795 event.[28] Turning to Africa, record low levels of the Nile from 1790 to 1797 suggest its extraordinary reach and true severity, in greatly reducing monsoonal rainfall on the Ethiopian highlands. In Egypt three successive years of exceptionally low floods led to famine and soaring wheat prices. This was followed in 1789 by the plague (called *Ta'oun Ismail Bey*), which lasted for five months. In 1791 and again in 1792, a slight drop below the long-term mean (and only two cubits or about one metre from the optimum water level) led to a severe famine. According to Antoune Zakry, people were forced to eat dead horses and donkeys and even children.[29] Another series of low floods in 1794, 1795 and 1796 led to a peasant revolt. This experience was comparable only with that of 1877 when the flood was two metres below average, leaving 62 per cent of Qena Province and 75 per cent of Girga Province unirrigated.[30]

Evidence from much of the rest of Africa is scanty. However, prolonged droughts in Natal and Zululand between 1789 and 1799 resulted in the *Mahlatule* famine.[31] This was the severest known in the written record to have affected southern Africa prior to an El Niño event of 1862.[32] The low rainfall shows up very clearly in dendrochronological records.[33] This ten-year drought began with an extraordinary record of cold sea conditions off the Cape. On 24 December 1789, as described in Chapter 13, the *Guardian*, carrying vital supplies to New South Wales, foundered on an iceberg right up near the Cape of Good Hope.[34]

The 'Great El Niño' and the French Revolution

In the normally temperate regions of western Europe, highly abnormal weather patterns were making themselves felt as early as 1787. There was an unusually cold winter in western Europe in 1787–1788, followed by a late and wet spring, and then a summer drought, resulting in the severe crop failures that helped to critically stimulate the explosive social pressures that culminated in the French Revolution.[35]

The extreme summer droughts and hailstorms of 1788 were decisive in their effects, as recorded by a peasant winegrower from near Meaux:

> In the year 1788, there was no winter, the spring was not favorable to crops, it was cold in the spring, the rye was not good, the wheat was quite good but the too great heat shrivelled the kernels so that the grain harvest was so small, hardly a sheaf or a peck, so that it was put off, but the wine harvest was very good and very good wines, gathered at the end of September, the wine was worth 25 livres after the harvest and the wheat 24 livres after the harvest, on July 13 there was a cloud of hail which began the other side of Paris and crossed all of France as far as Picardy, it did great damage, the hail weighed 8 livres, it cut down wheat and trees in its path, its course was two leagues wide by fifty long, some horses were killed.[36]

This hailstorm burst over a great part of central France, from Rouen in Normandy to as far as Toulouse in the south. Thomas Blaikie, who witnessed it, wrote of stones so monstrous that they killed hares and partridges and ripped branches from elm trees. The hailstorm wiped out budding vines in Alsace, Burgundy and the Loire, and laid waste to wheat fields in much of central France. Ripening fruit was damaged on the trees in the Midi and the Calvados regions. In the western province of the Beauce, the cereal crops had already survived one hailstorm on 29 May but succumbed to the second blow in July. Farmers south of Paris reported that, after July, the countryside had been reduced to an arid desert. These kinds of conditions led in late summer 1788 to serious rural unrest. Small-scale rural revolts took place in the areas worst affected by the summer droughts —

Provence, Hainault, Cambresis, Picardy, the area to the south of Paris, eastwards in Franche-Comte, around Lyon and Languedoc and westwards in Poitou and Brittany.[37]

To make matters worse, the prolonged drought was followed by the severest winter since 1709.[38] Rivers froze throughout the country and wolves were said to have descended from the Alps down into Languedoc. In the Tarn and the Ardèche, people were reduced to boiling tree bark to make gruel. Birds froze on the perches or fell from the sky. Watermills froze in their rivers and thus prevented the grinding of wheat for desperately needed flour. Snow lay on the ground as far south as Toulouse until late April. In January Count Honoré de Mirabeau visited Provence and wrote:

> Every scourge has been unloosed. Everywhere I have found men dead of cold and hunger, and that in the midst of wheat for lack of flour, all the mills being frozen.

Occasional thaws made the situation worse: the Loire burst its banks and flooded onto the streets of Blois and Tours.

All these winter disasters came on top of food shortages brought on by the droughts of the 1787 summer and the appalling harvests of the summer of 1788. Warm, dry spring-summers are usually favourable to grain in northern France and north-western Europe, but, at certain times, they can be disastrous. A spell of dry heat during the growth period, when the grain is still soft and moist and not yet hardened, can wither all hope of harvests in a few days. This is what happened in 1788, which had a good summer, early wine harvests and bad grain harvests. The wheat shrivelled, thus paving the way for the food crisis, the 'great fear', and the unrest of the hungry, when the time of the 'soudure' or bridging of the gap between harvests came in the spring of 1789. No one expressed this fear better than the poor woman with whom Arthur Young walked up a hill in Champagne on a July day in 1789:

> Her husband had a morsel of land, one cowe, and a poor litte horse, yet they had 42 lbs. of wheat and three chickens to pay as a quit-rent to one seigneur, and 168 lbs of oats, one chicken and one sou, to pay another, besides very heavy tailles and other taxes. She had 7 children, and the cow's milk helped to make the soup. It was said at present that something was to be done by some great folks for such poor ones, but she did not know who or how, but God send us better, *car les tailles et les droits nous écrasent* [for the cuts and taxes will ruin us].[39]

The price of bread had doubled between 1787 and October 1788. By midwinter 1788–1789, clergy estimated that a fifth of the population of Paris had become dependent on charitable relief of some sort. In the countryside landless labourers were especially badly affected. Exploitation of the dearth by grain traders and hoarders made the situation steadily worse. It was in this context that the French king requested communities throughout France to draw up '*cahiers*' [books] of complaints and grievances to be presented in Paris. During February to April 1789 over 25,000 *cahiers* were drawn up. From these we can assess not only the accumulation of long-term grievances, but also some idea of the intense dislocation of normal economic life that the extreme weather conditions had brought.

The excessive cold and food shortages of early 1789 drove increased poaching and stealing. There were regular attacks on grain transports both on road and river. Bakeries and granaries were also robbed. Rabbits, deer and other game were slaughtered, irrespective of ownership or regulation. Gamekeepers and other symbols of authority who opposed such actions were killed. The populace became accustomed to a level of resistance that soon developed into broader reaction and violent protest. Anger at the price of grain and bread in Paris soon found suitable targets for rioting and violence, particularly where the large population of quayside labourers were out of work because the Seine was still frozen in April. A number of pamphlets printed at this time proclaimed that the supply of bread should be the first object of the planned Estates-General: the very first duty of all true citizens was to 'tear from the jaws of death your co-citizens who groan at the very doors of your assemblies'.[40]

In the summer of 1789 much of rural France rose in revolt, while urban crowds rioted. How far the resulting course of revolution had its roots in the anomalous climatic situation of the period is open to debate. What is certain is that the part played by extreme weather events in bringing about social disturbance during the French Revolution cannot be neglected. It may be, as de Tocqueville put it, that had these responses to anomalous climatic events not occurred, 'the old social edifice would have none the less fallen everywhere, at one place sooner, at another later; only it would have fallen piece by piece, instead of collapsing in a single crash'.[41]

The Catalonian Revolt and the rest of Europe

Connections between an accumulation of unusual and extreme weather events and popular rebellion were by no means confined to France. In northern Mediterranean Spain, the cold winter of 1788–1789 was, if anything, even more unusual than in France. Here too, persistent summer droughts were followed by a winter of intense cold and heavy snowfall.[42] One observer wrote:

> Autumn this year was colder than normal ... and no one alive has ever experienced the weather so cold in El Prat. It was extraordinary, both what was observed and the effects it caused ... on the 30th and 31st December the wash of the waves on the beach froze which has also never been seen or heard of before. Likewise it was observed that the water froze in the washbasins in the cells where the nuns slept at the Religious Order of Compassion ... the river channels froze and the carriages passed over the ice without breaking it.[43]

Between August 1788 and February 1789, cereal prices in Barcelona rose by 50 per cent.[44] Between February and March 1789 there was a revolt in the city, *Rebomboris de Pa*, when the population set fire to the municipal stores and ovens. The authorities attempted to pacify the angry crowds by handing out provisions and offering supplies at reasonable prices. The privileged classes also provided money and contributions in kind to pacify the underprivileged, whilst the military and police authorities stepped back to allow events to run their course.

The authorities then took refuge in the two fortresses that controlled the city and put up powerful defences in case events got out of control. Despite these measures, chaotic riots ensued and in the aftermath six people were executed.[45] Similar riots took place in other parts of Catalonia when the poor outlook for the 1789 harvest became clear and profiteers and hoarders made their appearance. Revolts and emergency actions by municipal authorities took place both on the coast and inland with documentary reports being made in cities such as Vic, Mataro and Tortosa.[46] The fact that these social responses to cold and crop failure did not lead to the same degree of social turmoil and rebellion as in France should not disguise the fact that they were highly unusual.

Even where there were no revolutions, the extraordinary weather patterns were commented upon. In the English winter of 1790, Parson Woodforde's diary tells of unusually high temperatures and summer-like weather in January and February of 1790.[47]

One of the advantages in trying to understand European history in terms of the succession of prior climatic stresses is that it provides an environmental context for events of the major revolutions, rather than analysing them as purely social phenomena. As de Tocqueville commented, 'The French Revolution will only be the darkness of night to those who see it in isolation; only the times which preceded it will give the light to illuminate it'.[48]

The whole social edifice of *Ancien Regime* France collapsed at a single blow, and the fact that this was in the midst of one of the worst El Niño episodes of the millennium is something that should be taken into account in examining this history.

The Great El Niño in North America

High winter temperatures were also experienced in North America in the early 1790s. Contemporary observers commented that horse herds had expanded greatly in numbers. These facilitated expansion and migrations by the Cree, Assiniboine, Blackfoot and Gros Ventre peoples in parts of Washington, Montana and Wyoming. The first three years of the 1790s were very warm and dry on the northern Great Plains. Fur traders in the region repeatedly remarked on how warm and snowless those winters were. High temperatures were accompanied by high rainfall events. On 13 January 1791 the first of a series of very heavy thunderstorms in the region is recorded, in Saskatchewan. This produced hardships for bison hunters, but horse herds multiplied. Hostilities among native groups were rare in those years. That warm episode ended abruptly in 1793–1794. At the end of the El Niño event, the return of cold winters provoked wars between the Indian tribes as conditions deteriorated for them and their horses. Horse herds were decimated, and warfare reached a climax in the next year.[49]

Further south, the conditions also gave rise to heavy rainfall and high temperatures in the United States, which brought an inexorable rise in the mosquito population. As a result, by 1793 the conditions were ideal for the spread of mosquito-borne diseases. On 19 August 1793, Dr Benjamin Rush, a doctor in the relatively northerly city of Philadelphia, noted his first cases of yellow fever.[50] This is a disease normally spread in tropical America by *Aedes aegypti*, a mosquito with a pronounced tropical range. By October 1793 the epidemic had killed over 5000 people in Philadelphia alone, and the epidemic was only ended by a severe frost in November, which killed the mosquitoes. It seems that the epidemic had spread from the French colony of Saint Domingue (now Haiti), where a slave rebellion had been aggravated by prolonged bad weather and crop failure in Europe. Refugees from the rebellion carried the yellow fever with them to the east coast ports of the United States, where the aberrantly high mosquito population allowed the disease to flourish. The political consequences of this disease event were far reaching and resulted, not least, in the French losing Haiti and relinquishing most of their American colonies.

Other diseases flourished in North America during the period of the Great El Niño. This was particularly the case with influenza, epidemics of which had previously rarely affected the North American mainland. An epidemic spread from Georgia in the southern United States to Nova Scotia, Canada, between September and December 1789, and again in spring 1790. Its most famous victim was George Washington. Dr Warren recorded that 'at New York, as far as I can learn, its appearance was somewhat later than here, and our beloved President Washington is but now on the recovery from a very severe and dangerous attack of it in that city'.[51] Thomas Pettigrew observed that 'the summer preceding the fall disease, was remarkably hot ... the last winter was uncommonly mild and rainy. The diseases of that season numerous, [including] both synocha and typhus'.[52] Certainly the very hot summers and mild winters which characterise El Niño conditions in much of North America appear to have encouraged the spread of epidemics in several different diseases, and not least in 1788–1794.

Conclusions

While further archival research is needed to characterise the 1789–1795 event more fully, the evidence of an intense and prolonged global impact already suggests that it may have been among the most severe in the available written record. The data from this well-documented period can throw light on exceptional global weather patterns. The correlation between a weak phase in the North Atlantic Oscillation (NAO) and an extreme El Niño episode elsewhere suggests a pattern tying NAO and ENSO together via the South Asian monsoon. There were already well-established strong correlations between inter-annual variability in the North Atlantic Oscillation and Indian summer monsoon rainfall, specifically between summer monsoon rainfall and the NAO of the preceding year in January.

The early stages of the Great El Niño event were observable in southern India *more than a year before* the warm current was recorded along the Peruvian coast, the usual key indicator of an El Niño phenomenon. This was also the case for the major El Niño event of the seventeenth century in 1685–1688. Moreover, it appears that in both episodes a presumed weak phase of the North Atlantic Oscillation led to a very cold winter, high pressure over Europe, a cold wet spring and a summer drought preceding later monsoonal blocking and drought in South and South-East Asia.

We are only now realising that this sequence of events and phasing of activity seem to form a global pattern. The mechanisms of the North Atlantic Oscillation, the Asian monsoon and the Pacific El Niño-Southern Oscillation appear to be closely articulated in the case of the severest El Niño episodes.

The connection ('teleconnection') between the failure of Indian monsoons and Australian droughts was noted in Australia by Charles Todd as early as 1888.[53] In 1896, the New South Wales Government Observer, H. C. Russell, carefully documented the coincidences between India and Australia from 1789 to 1886, and showed the correlations to be strongest in the case of the worst episodes (see Table 3).[54]

TABLE 3
COINCIDENCE OF AUSTRALIAN AND INDIAN DROUGHTS
(AFTER RUSSELL)

Australian droughts	Indian droughts
1789–1791	1790–1792
1793	
1797	
1798–1800	
1802–1804	1802–1804
1808–1815	1808–1813
1818–1821	
1824	1824–1825
1827–1829	1828
1833	1832–1833
1837–1839	1837–1839
1842–1843	
1846–1847	
1849–1852	
1855	
1857–1859	1856–1858
1861–1862	
1865–1869	1865–1866
1872	
1875–1877	1875–1877
1880–1881	
1884–1886	1884

The developmental sequence of the 1788–1795 El Niño may provide a model to compare further with other subsequent very severe El Niño events since that time, for example: 1895–1902, 1982–1983 and 1997–1998. It may also serve as a template for understanding much earlier events in the historical record where the data is more sketchy. Such data would give retrospective comfort to the observant Arthur Phillip, that his hunch was right. The events do not directly correlate, but rather form a pattern that may take several years to work sequentially through the global weather and ocean system.[55] 'Persistent' or extended El Niños such as that of 1788–1795 suggest a sequence for these patterns.

Richard Grove

Richard Grove is Research Director of the Centre for World Environmental History at the University of Sussex, Visiting Professor at Jawarhalal Nehru University, New Delhi and Senior Fellow, Resource Management in Asia-Pacific Program, Research School of Pacific and Asian Studies, Australian National University. His books include *Green*

Imperialism (1995), *Nature and the Orient* (1997) and *Ecology, Climate and Empire: Studies in Colonial Environmental History* (1997). In 2000, he co-edited (with John Chappell) *El Niño: History and Crisis*. He is foundation editor of the international journals *Environment and History* (Cambridge, UK) and *International Journal of Environmental History and Environmental Justice* (Pondicherry, India).

12

Climate change and human health

looking backwards, planning forward

Tony McMichael

Over the past two decades, it has become clear that *Homo sapiens*, in aggregate, is now exerting huge pressure on the world's natural environment. This explains the recent acceleration of global climate change, depletion of stratospheric ozone, increasing loss of biodiversity, impairment of food-producing systems on land and in the oceans, and a worldwide draw-down of freshwater stocks. Global environmental changes reflect a rapid increase in human domination of the biosphere over the past century, as human numbers have increased and as human economic activities have intensified by more than an order of magnitude.[1]

We appear to have passed certain critical points. In consequence, our collective 'ecological footprint' has become greater than the biosphere can accommodate. Indeed, one recent detailed environmental analysis shows that humankind has been operating in ecological deficit since the 1970s: that is, we now consume and deplete Earth's natural environmental capital at a greater rate than it is being replenished and restored.[2] Most other similar assessments broadly concur.

In 2001, the Third Assessment Report of the United Nations Inter-governmental Panel on Climate Change (IPCC) documented a persuasive pattern of recent changes observed in various physical and simple biological systems.[3] This included glacier retreat, sea ice diminution, earlier bird nesting, earlier bud burst and flowering, and changes in insect migration times. The IPCC pointed out that this pattern, viewed overall, points clearly to the incipient impact of warming conditions around the world.

Where and when, therefore, might we see health impacts of climate change in human populations? The answer to that question is complex. First, most health outcomes are multi-causal, and it is likely that many of those non-climate causal factors are also changing over time. It is therefore difficult to disentangle climatic and non-climatic influences. Second, climate change affects local environments differently, according to characteristics of local geography. Further, the vulnerability of each human population varies as a function of locality, level of material resources, technological assets and type of governance.

There is some — though inconclusive — published evidence that we may now be witnessing some early impacts of climate change on infectious diseases. For example, tick-borne (viral) encephalitis in Sweden appears to have increased in response to a succession of warmer winters over the past two decades.[4] The tick, however, has a complex three-stage life-cycle, and the relationship of tick development to temperature and rainfall is complex, entailing a particular sensitivity of the early, larval stage. There is also some suggestive evidence that malaria may be responding to warmer temperatures in east African highlands, by moving to higher altitudes.[5] This complex issue, however, remains unresolved.[6] Meanwhile, intensification of the El Niño-Southern Oscillation (ENSO) over the past quarter century, itself a presumed consequence of global climate change, has been accompanied by a strengthening inter-annual association with diarrhoeal disease in Bangladesh.[7]

It is important to note that these various health impacts mostly do not entail novel outcomes (in contrast to the surprise appearance, in recent decades, of HIV/AIDS and human 'mad cow disease'). Instead, they entail changes in the frequency or severity of familiar health risks, such as the impacts of floods, storms and fires; the mortality toll of heatwaves; the range and seasonality of infectious diseases; changes in local agricultural productivity and the nutritional consequences; the health impacts of altered supplies of fresh water; and the health consequences of economic dislocation and population displacement.

Of course, the health prospects of climate change are not all bad. For example, milder winters would reduce the seasonal winter time mortality peak in temperate countries, whereas a further increase in temperatures in currently hot regions might impair mosquito survival. Overall, however, scientists have consistently assessed that most climate change health impacts would be adverse.[8]

Studies in Australia will enrich the information base for estimating the impacts of changes in climatic means and variability. The dependence of Ross River Virus disease on inter-annual climate variation and on birds as vectors is becoming clearer.[9] Various other studies have been reported on the climatic influences on Murray Valley Encephalitis and on mortality from thermal stress. New information is now available on the relationship between short-term variations in temperature and the occurrence of food-poisoning by salmonella and campylobacter in Australia.[10] The Commonwealth government, in 2002, commissioned a first formal assessment of the health impacts of projected scenarios of climate change in Australia over the coming decades, which has yet to be published.

Meanwhile, as average surface temperatures gradually rise, it is likely that we will face an increase in climatic variability, including extreme weather events.[11] Indeed, many scientists now consider that human health and safety are more endangered by extreme and anomalous weather events than by the change in average climate conditions.

Climate and human health: an ancient link

That human health and disease reflect climatic conditions is an idea that almost certainly predates written history. The Greek physician Hippocrates, around 400

BC, wrote that physicians should have 'due regard to the seasons of the year, and the diseases which they produce, and to the states of the wind peculiar to each country and the qualities of its waters'. He instructed them to observe 'the waters which people use, whether they be marshy and soft, or hard and running from elevated and rocky situations, and then if saltish and unfit for cooking' and to note also 'the localities of towns, and of the surrounding country, whether they are low or high, hot or cold, wet or dry ... and of the diet and regimen of the inhabitants'.[12]

Since Hippocrates' time, there is evidence of many climatic disasters affecting communities and populations around the world. These caused starvation, infectious disease, social collapse and, sometimes, the disappearance of whole populations. The famous Justinian Plague, which decimated much of the Roman Empire in the mid-fifth century AD, followed a decade of low temperatures and bad weather in the Mediterranean region. From research elsewhere, there is evidence that the infected rodents that harbour the plague bacterium become more mobile during times of unusual weather. This may also have contributed to the timing of the outbreak of the Black Death in Europe in the mid-fourteenth century, via prolonged unusual weather patterns and regional drying in the Asian Steppes. Some commentators have suggested that this facilitated the contact of infected rats with long-distance traders and Tartar raiders.[13]

One dramatic example is afforded by the somewhat mysterious demise of the two Viking settlements in southern and western Greenland in the fourteenth and fifteenth centuries, as temperatures around Europe began to fall with the onset of the Little Ice Age.[14] Culturally conservative and livestock-dependent settlements were established during the tenth century AD — partly in response to some devious advertising by Eric the Red who named the largely ice-bound landmass 'Greenland'. Settlement occurred, fortuitously, early in the Medieval Warm Period. The settlers, however, could not cope with the progressively colder climates of the late Middle Ages, from approximately the late thirteenth century. Food production declined and food importation became more difficult as sea ice persisted longer. Trade with Europe became more difficult; a letter from the Pope took over six months to be delivered! Viking settlements in Greenland eventually died out or were abandoned. The reason for their demise remains undetermined.

There are numerous historical accounts of acute famines occurring in response to climatic fluctuations.[15] In pre-industrial Europe, food supplies were marginal, and the mass of people survived on monotonous diets of vegetables, grain gruel and bread. A particularly dramatic example in Europe was the great medieval famine of 1315–1317. Climatic conditions were deteriorating, and the cold and soggy weather led to widespread crop failures, food price rises, hunger and death. Social unrest increased, robberies multiplied, and bands of desperate peasants swarmed over the countryside. Animal diseases proliferated, contributing to the death of more than half the sheep and oxen in Europe. These climatic setbacks, and the Black Death that followed thirty years later, may have contributed to the weakening and dissolution of feudalism in Europe.

Recurrent famines repeatedly culled the populations. In Tuscany, for example, famine occurred in more than 100 years between the fourteenth and eighteenth centuries. In China between around 100 BC and 1900 AD, where the rural diet of vegetables and rice provided an estimated 98 per cent of caloric intake, famines that involved at least one province occurred in more than 90 per cent of years.

Extremes of weather, often as manifestations of some of the world's great regional and global climatic oscillations — such as the well-known El Niño-Southern Oscillation (ENSO) — have frequently caused devastation in human societies and local populations. The potency of these extreme events is well illustrated by the great El Niño event of 1877–1878, one of the strongest such episodes on record.[16] An estimated ten million people perished in northern China and eight million in India due to famines. Up to one million people also died in north-east Brazil, and droughts caused food shortages in famine throughout Africa. Meanwhile, Tahiti was struck by a 'terrible hurricane'. In the United States, large and persistent temperature and precipitation anomalies were recorded from mid-1877 to mid-1878. In the unusually mild and humid summer of 1878, one of the most severe outbreaks of mosquito-borne yellow fever affected the southern United States. The economic and human toll was enormous: in the city of Memphis, Tennessee, up to 20,000 people died out of approximately 100,000 cases of the disease.

An even more severe El Niño event had occurred in the period 1788–1791.[17] This event caused massive famine and starvation in India. The hapless, newly arrived settlers (mostly convicts) in Sydney Cove, Australia, were faced by water shortages, dryness and heat; their crops wilted. Weather patterns as far afield as Europe were disrupted.

Similarly widespread impacts have accompanied more recent El Niño events. In 1982–1983, the floods on the western coast of South America caused a tripling in the incidence of acute diarrhoeal diseases and acute respiratory diseases in Bolivia and in Peru, and great increases in infant mortality due to malnutrition. Flooding in the Horn of Africa associated with the El Niño of 1997–1998 was linked with an upsurge of cholera deaths due to disrupted sanitation and contaminated water supplies. Meanwhile, the burning of millions of hectares of forest in Indonesia was amplified by drought due to that El Niño episode. The resultant smoke haze blanketed the South-East Asian region and northern Australia, and caused sharp rises in respiratory illness in the worst affected areas.

The infectious diseases caused by the Hantaan viruses provide another recent example. These viral diseases have been known in rural Asia for centuries, but they have now spread elsewhere because of economic and ecological changes that have increased contact between rodents and humans. These rodents harbour a virus that can be transmitted to humans via dried excreta. In 1993 this type of viral disease was first detected in the United States. The prolonged El Niño event of the early 1990s initially caused drought conditions in the south-west United States. This led to a decline in vegetation and in animal populations, including the natural predators (owls, snakes and coyotes) of the deer mouse. When heavy rains occurred in 1993, inducing proliferation of grasshoppers and piñon nuts, there was an unchecked tenfold expansion in the deer-mouse population, resulting in hantavirus pulmonary disease in humans — a disease that was first described at that time. This disease has subsequently spread to many contiguous states, western Canada, and much of Latin America (especially Argentina).[18]

Potential health impacts of global climate change

There is as yet little direct empirical evidence of how decadal-length climate change affects human health. While that evidence base grows, various researchers continue their attempts to model how future climate scenarios, given certain 'future worlds' and their associated emissions of greenhouse gases, might affect human population health. Several generic categories of health impacts due to climate change may be distinguished: first, those that arise from relatively direct impacts of alterations in temperature, precipitation and extreme weather events; second, those that occur in response to climate-induced changes in ecological processes and systems; and third, those that result from the economic and demographic dislocation of human communities.

Direct effects

The more direct and immediate impacts include those due to changes in exposure to very cold and very hot weather extremes and those due to increases in extreme weather events (floods, tropical cyclones, storm-surges and droughts). Climate change would also directly increase the production of certain air pollutants, such as tropospheric ozone (the formation of which is affected by both temperature and level of sunlight) and various aeroallergens (spores and moulds) that contribute to asthma, hay fever and other allergic disorders.

Each population displays a characteristic pattern and level of daily deaths (and hospitalisations) in relation to short-term prevailing temperature.[19] Typically, death rates increase both at times of greater heat and greater cold. In temperate countries, climate change would result in a decrease in winter mortality because of less severe winters, and this may offset — or more than offset — the increases in summer heat-related mortality.[20] The net balance of future changes in hot and cold impacts remains uncertain — and besides, this balance varies between different geographic regions, between different levels of economic development, and between urban and rural populations.

The extent to which the regional frequency of extreme weather events will alter because of climate change remains somewhat uncertain. Climatologists, however, are becoming increasingly confident of their modelled regional predictions, as the downscaling of their climate models improves. Further, extreme weather events appear to have increased in tempo during the 1990s, with commensurate impacts on human life and limb.[21]

Nevertheless, despite the potentially great impact of such events on deaths, injuries and consequent diseases (infections, malnutrition and mental health disorders), estimating the future profile of health impacts is at best an indicative exercise. Hurricane Mitch, centred in the Honduras in 1998, was followed by tens of thousands of new cases of malaria, cholera and dengue fever.

Indirect effects (especially infectious diseases and food yields)

The indirect, second category of health impact refers particularly to changes in the transmission patterns of infectious diseases and changes in regional food-producing systems. These impacts are, in essence, a result of ecologically mediated mechanisms.

In the longer term, it is probable that these indirect impacts on human health would have greater magnitude than the more direct impacts.[22]

For vector-borne infectious diseases, the geographic range and abundance of vector organisms (and in some cases, their intermediate hosts) are affected by various meteorological factors such as temperature, precipitation, humidity, surface water and wind; and biotic factors like vegetation, host species, predators, competitors, parasites and human interventions. Further, the rate of maturation of pathogens within the vector organism is typically sensitive to temperature, in curvilinear fashion, such that a particular increment of warming across different parts of the temperature range yields very different increases in transmissibility.

We can expect that a shift in background climate conditions would affect the spatial and seasonal patterns of the potential transmission of various vector-borne infectious diseases. These include malaria, dengue fever, the various types of viral encephalitis (brain infection), schistosomiasis (or 'bilharzia', spread by water snails), leishmaniasis (spread by sand flies in South America and around the Mediterranean coast), onchocerciasis (West African 'river blindness', spread by black flies) and yellow fever (also spread by the *Aedes aegypti* mosquito). The key phrase in this discussion is 'potential transmission'. It is important to estimate how the intrinsic disease-transmission properties of the world might alter in response to climate change. Indeed, such research is in the classic tradition of experimental science, which seeks to hold everything else constant while estimating the effect of varying just one key factor. Such estimations can be made with climate change as that one key factor, the putative determinant of the future transmissibility of infectious disease. Nevertheless, we know that the actual transmission of diseases such as malaria is, and will be, much affected by economic and social conditions, and by the robustness of public health defences. Hence we must also develop models that incorporate other reasonably foreseeable contextual changes. This is a tall order, but we have no other way of exploring the likely future risks of climate change to human health.

The main anticipated impact of climate change on the potential transmission of vector-borne diseases would be in tropical regions. In general, populations on the margins of endemic areas in tropical and sub-tropical countries would be most likely to experience an increase in transmission. Indeed, malaria is currently resurgent in many countries where, several decades ago, it had been greatly reduced with vector control measures. Climate change may compound that process — as it may also do with dengue fever. That disease, like malaria, has also increased substantially in recent decades, particularly in Central and South America. This appears to reflect a combination of increasing population mobility, urbanisation, poverty and regional warming, and a slackening of mosquito control programs. Meanwhile, in temperate zones, climate change may also affect diseases such as tick-borne viral encephalitis (which occurs in parts of western Europe, Russia and Scandinavia), Ross River Virus disease (which occurs in Australia), and Lyme disease (which occurs particularly in north-east United States and parts of Europe).

Estimation of future climate-induced changes in the potential transmission of malaria best illustrates the uses, and limitations, of computerised mathematical modelling. Considered overall, such models project that under conditions of standard scenarios of climate change, the geographic range of potential malaria transmission would undergo a net, albeit small, expansion. One detailed and plausible modelling study has estimated that an additional 3 to 4 per cent of the world's people (that is several hundred million people) would live in regions of potential transmission of malaria in the second half of the twenty-first century.[23] Modelled projections have also been made for dengue fever and schistosomiasis. Interestingly, in some regions global warming may actually diminish the geographic range of schistosomiasis transmissibility because warmer water would reduce the viability of certain water snail vector species.

These predictive models, however, have their limitations.[24] They are limited by incomplete information about the biological and ecological processes that they attempt to model; by difficulties in finding relevant historical data against which to validate them; and by their inability to simulate fully the rich tapestry of social, economic, demographic, behavioural and technological changes in human societies that will, of course, accompany future climate change. Therefore they perform better in modelling the risks of potential climate-permitted transmission than in modelling the rates of actual transmission. Nevertheless, they represent the best that science can do to foresee the direction and approximate magnitude of the climate-induced shifts in risk that future populations will face. Recent efforts have sought to develop better mathematical models for making such scenario-based projections.

Climate change would also affect the transmission of various water-borne infectious diseases. This would occur via several mechanisms, including intensified rainfall leading to flooding and contamination of drinking water supplies. Bacterial water-borne illnesses such as gastroenteritis due to coliform bacteria, giardiasis, and cholera may thus be affected; so, too, would various protozoal water-borne infectious diseases such as cryptosporidiosis. Similarly, warmer temperatures would tend to increase the summer seasonal peaks of food-borne bacterial enteric infections, such as those caused by salmonella and campylobacter. In low-income countries, with poor hygiene and resources, rates of serious diarrhoeal disease in children would rise. The sensitivity of child diarrhoea to variations in climatic conditions has been well demonstrated in studies in Lima, Peru, and in Fiji.[25]

Another fundamentally important potential impact on human health would arise from any downturn in food availability. Climate change is likely to affect crop yields, livestock health and resultant food products, and fisheries. However, the processes are generally complex — and often entail seemingly stochastic aspects (for example, the occurrence of new infectious diseases in plants and animals). Cereal grains account for around 60 per cent of world food energy — both via direct consumption and via the feeding of grains to livestock for meat production. Therefore, not surprisingly, various modelling studies have made estimates of the impacts of climate change upon cereal grain yields — allowing for the expected 'fertilisation' effect of the increasing concentration of atmospheric carbon dioxide. Globally, a slight downturn in grain production appears likely, especially in the latter half of the twenty-first century.[26] This downturn would be greater in the already food insecure regions in South Asia, parts of Africa and Central America, increasing the number of malnourished people in the world (currently an estimated 830 million) by at least several per cent overall — with higher percentage increases in sub-Saharan Africa and South Asia. World cereal grain yields have become a little more unstable during the 1990s, displaying increased inter-annual variability. Could this be (partly) due to changing climatic conditions?

Effects of economic, social and demographic disruption

The third category of health impact is somewhat speculative and difficult to model quantitatively — but it can be confidently predicted, given the experience of displaced or economically disrupted populations elsewhere. Refugee populations experience a range of risks to health, including mental health problems, malnutrition, infectious diseases, and the physical hazards of new and improvised living environments. Several small island states are registering growing concern about sea level rise, and this concern itself may be jeopardising well-being and mental health. Economic downturns and unemployment are conducive to a similar range of risks to health. Further, such disturbances and displacements of populations can create conflict situations, both internally within the affected community and in its external relations with other communities.

These anticipated climate change-related health consequences of economic disruptions, of rising sea level, and of population displacement for reasons of physical hazard, land loss, unemployment and civil strife, may not become evident for several decades. The usual difficulty for researchers — mentioned previously in this chapter — is in attributing the causation of such outcomes in situations where there are many inter-related influences involved. It will therefore be particularly difficult to detect incipient climate 'signals' via these more indirect pathways, against the considerable background noise.

Population vulnerability and adaptive responses

The magnitude of the impact of climate change upon health depends both on the extent of that change and on characteristics of the target population. The population's 'vulnerability' is a function both of the sensitivity of the exposure response function in that population and of the population's adaptive (that is, impact lessening) capacity.[27] That adaptive capacity depends on factors such as population density, level of economic development, food availability, local environmental conditions, pre-existing health status, and the quality and availability of public health care. It also depends on various structural and politically determined characteristics, including socio-cultural rigidity, international connectedness and political flexibility.[28]

The actual level of risk to the population's health is therefore a function of that vulnerability and the level of exposure to climatic or associated environmental factors. Because we are already committed to some degree of global climate change over coming decades, despite any mitigating actions we might take now, it is important to begin assessing, for any particular population, both its vulnerability and its adaptation options.

Conclusion

The inter-connections between human demography, economic activity, environmental conditions and the vitality of the biosphere are becoming clearer. Meanwhile, discussions and formulations that extend across disciplinary boundaries are, at last, becoming more enlightened. We are recognising that human societies and their well-being and health are beholden to the sustained integrity of ecological relationships and environmental processes.[29]

One of the enduring legacies of the differentiated structure of modern western science, as it evolved during the past two centuries, is that discipline-bound ships kept passing in the night. The various physical, biological and social sciences have retained, indeed reinforced, a separateness of topic, theory and idiom. This divided approach to knowledge, discussed sympathetically and engagingly by E. O. Wilson in his book *Consilience*, has both strengths and weaknesses.[30] The strengths come from specialisation and the power of reductionist thinking. The weaknesses come from tunnel vision and disconnectedness.

There is a need for a more integrated type of knowledge. Encouragingly, there is a recognition that various of the most pressing environmental and social problems are large-scale, multi-sectoral and truly complex. In such a world, there is special need for interdisciplinary, indeed, transdisciplinary, approaches — that is, approaches that harness the ideas and energies of assorted disciplines within a synergistic team. This has been a distinctive feature of much of the past decade's scientific research in the field of global climate change and its actual and potential impacts.

Tony McMichael

Professor Tony McMichael is Director of the National Centre for Epidemiology and Population Health, Australian National University. Over the past several decades he has done epidemiological studies of occupational diseases, dietary influences on non-communicable diseases, socioeconomic health differentials, and environmental health problems. He was, from 1994 until 2001, Professor of Epidemiology at the London School of Hygiene and Tropical Medicine. While there, he established the Centre on Globalisation, Environmental Change and Health. Professor McMichael has interests in the population health consequences of global environmental change and ecosystem disruption. Since 1993 he has coordinated the assessment of health impacts for the United Nations Intergovernmental Panel on Climate Change. His book, *Human Frontiers, Environments and Disease: Past Patterns, Uncertain Futures*, was published in 2001 by Cambridge University Press.

Advertisement for Hills hoists printed in *Australian Women's Weekly*, 28 November 1956, p. 55

Winds Rising

The Hills hoist is a suburban Australian icon. Although popularly known as 'Hill's hoists', the rotary clothes line has a longer history than its production in post-War Adelaide. Melbourne blacksmith Gilbert Toyne advertised Australia's first rotary clothesline in the 1920s, although the design was first patented in America in the 1890s.

The Hills company, established by Lance Hill, grew from a backyard production line to become a major industry. Initially, Lance built a clothes line out of old bits of tubing for his wife who was frustrated with her existing wall-mounted line. Word spread quickly around the neighbourhood and when Lance overheard two local ladies speaking about Mrs Hill's wonderful new acquisition, he immediately set about producing more hoists. Although the design was not new, the hoist mechanism was, and it was this that Lance patented in 1956. From a fortune built on drying sheets in the sun and wind, Hills Industries grew into an internationally renowned manufacturer.

Display of Hills hoists in the Nation gallery of the National Museum of Australia

PHOTOGRAPH BY BRENDAN BELL

13

The roaring forties

Tom Griffiths

Moving air is a powerful element. We inhabit a swirling, gaseous soup that is animated by the heat of the sun and the spinning of the earth. 'Somehow', writes Jan DeBlieu in her meditation on *Wind*, 'out in the elements, the wisdom of science falls a bit short. It is easier to believe that wind is the roaring breath of a serpent who lives just over the horizon'.[1]

Airs, waters, places

Pioneer Australian meteorologist, W. S. Jevons, considered wind to be 'the great cause of all weather', 'the Great Locomotive of the Globe'. 'Winds are not local and fortuitous occurrences', he wrote in 1859, 'they are vast streams of air ceaselessly traversing the surface of the earth from pole to equator, from one continent or ocean to another'. He yearned to see the winds from above, from out in space, a whole hemisphere at a glance, knowing he would see it 'covered by innumerable streaks of wind, all curved and involved together in the most complicated manner'; he would then 'learn more of meteorology than all meteorologists have done'.[2]

Although Jevons wanted to see the global 'edifice of winds' at a glance, he knew Australians must painstakingly build up a constellation of detailed local knowledges. Henry Hunt, who was appointed the first Commonwealth Meteorologist in 1907, gained his fame in 1894 with a prize-winning essay on Sydney's 'southerly buster', which was regarded as 'perhaps the most remarkable of the "squall" winds which are found in various parts of the earth'. The 'southerly buster', or 'brickfielder' as it was also known, brought oppressive heat to a sudden end with a dramatic afternoon squall of wind that plunged the temperature and carried a whirling cloud of red dust 'derived from the pulverised ironstone of the streets'.[3]

Some of the earliest Australian meteorological speculations were medical. Doctors brought to colonial Australia a Hippocratic concern with seasonal and environmental influences on health and disease. They worried especially about the winds. Hippocrates' treatise, *Airs, Waters, and Places*, written in the fifth century BC, urged physicians who arrived in a new locality to map the airs, smell their quality, and find their source.[4] The new land was to be subjected to a respiratory examination. Medicine was a topographic and meteorological science, and in strange terrain it became a discourse of settlement, 'another way of knowing the country'.[5]

Colonial doctors monitored the effects of alien airs on the transplanted bodies of Britons; they tried to diagnose the local winds and waters. Warwick Anderson, in his book *The Cultivation of Whiteness*, has recalled the power of this colonial medical geography before the rise of contagionist doctrines and germ theory in the 1880s, a world in which ill health was often construed as environmental imbalance. Colonists, writes Anderson, 'felt their bodies move along with the strange elasticity of the air; their spirits shifted with the direction of the winds'; they used 'their own bodily sensations, their feelings of comfort or unease, to judge whether the land they coveted was a properly British territory'. Hot winds from the inland constituted an assault on the body's balance, they could provoke lassitude and depression, they were dangerous destabilisers of morality and morale, they could be degenerative. They emanated from the little-known heart of the continent, from the primitive desert. As David Walker explains in Chapter 8, geography, history and health were fatally related. Racial and bodily integrity could be threatened or ameliorated by emigration, and the confident material possession of Australia by Europeans could be unsettled biologically by sudden, hot winds from the desert — or, for that matter, by an Antarctic blast from the ocean.[6]

Australian airs were also touted for their beneficial effects. A warm, dry climate and the bracing purity of the air were promoted as providing an effective cure for tuberculosis, the major cause of death in Great Britain in the nineteenth century. It was not just the destination that was thought to improve the health of the consumptive immigrant, but also the passage by sea. A nineteenth-century medical cult of sea cures, and the promotion of 'the ocean as health resort', lured ailing immigrants to Australia's southern shores, especially Victoria, whose sudden changes of temperature were thought to be stimulating. But Australian death rates for tuberculosis were still rising while those in Britain had begun to fall, and consumptive patients who were induced to undertake a stressful, crowded sea voyage in tempestuous high latitudes often found that they had chosen death rather than Australia.[7] The southern seas that were the gateway to the colonies, one doctor warned, stretched 'to the South Pole in almost unbroken extent', and these same seas gnawed the coastline of Australia and governed its climate. If they must come to Australia for the cure, the Melbourne ophthalmologist James Barrett told prospective immigrants, then '*in all cases they must remember that their place of safety lies north of the Great Dividing Range*', away from the punishing winds of the polar ocean.[8]

The Albatross Latitudes

'The roaring forties' were the earliest and most famous southern winds in the European imagination. They shaped the European discovery of Australia, influenced the foundation of British settlement in the Pacific, and brought Australia's nineteenth-century immigrants to their new home. Encountering and surviving these wild westerlies on the Southern Ocean was a rite of passage for all newcomers to Australia, a common dimension of their new identity.

Europeans who happened upon Australia from the seventeenth century travelled in ships that first voyaged confidently across latitude and then easted anxiously within it. Latitude could be read accurately from the sun, while longitude remained

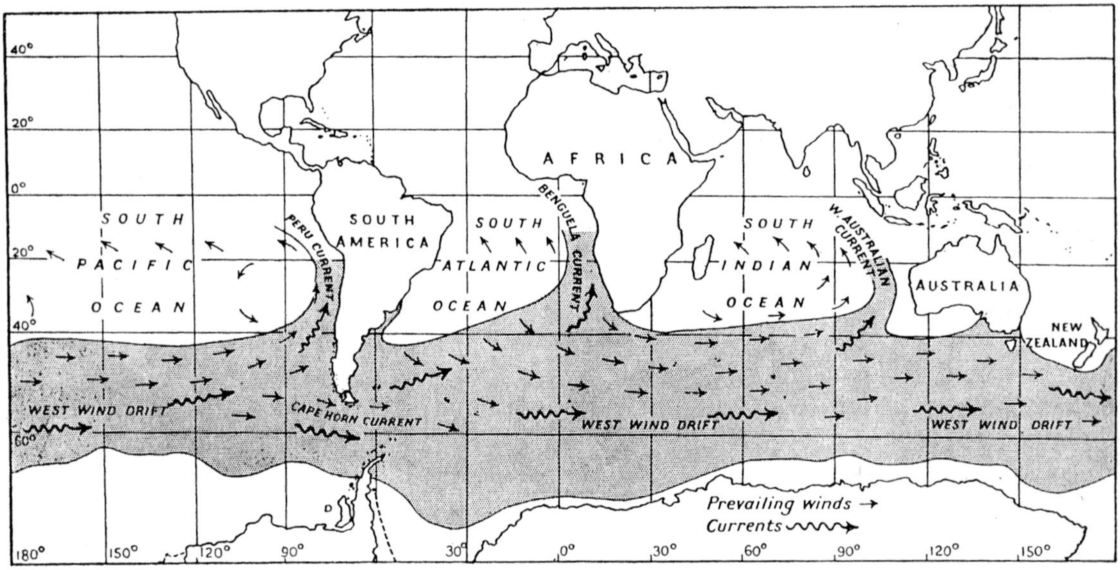

THE ALBATROSS LATITUDES

Map reproduced from William Jameson, *The Wandering Albatross*, Rupert Hart-Davis, London, 1958.

the chief problem in navigation until the late eighteenth century. Sailing south in the Atlantic, there was a known gradation of airs. There were the low pressure doldrums at the equator, where ships could be trapped in a sticky calm, then the steady, reliable trade winds to the north and south. Next there was the unruffled air of 'the horse latitudes' around the Tropics of Capricorn and Cancer, the latitude of many of the world's deserts. The horse latitudes took their name from the fact that it was here that the crews of becalmed sailing ships threw horses overboard when drinking water became scarce. South of the 35th parallel — which was about the Cape of Good Hope, the end of known land — ships began to edge into the swirling atmospheric engine of the Earth. Here crews and passengers found themselves engulfed by an elemental force that was both exhilarating and awesome. The Southern Ocean, observed one French historian, was above all, 'une realité pour nos oreilles' [a world for our ears]. Latitude was manifest in the violent music of the winds: in the forties they roared, in the fifties they screamed, and in the sixties they whistled.[9]

The prevailing westerlies of these higher southern latitudes encircle the bottom of the globe and isolate Antarctica. They boom and broil unimpeded by land. The break-up of Gondwana created their vast playground. The Southern Ocean is a pure realm of water and wind, an extensive belt of the earth's surface where these two fluids encounter and animate one another without interruption. The dance of air and sea is vigorous; each reaches for and agitates the other. Here an ocean swell greater than three metres is usual, and waves of 10–11 metres or more are a regular occurrence, with swells almost as high. A ship might be lost from sight between waves. Clippers running before a gale could travel over 300 miles in a day. Mean monthly wind speeds between the latitudes of 40 and 60 degrees south are 20–45 kilometres per hour, and gales and storms are frequent.[10] Torrent and current combine to become one inexorable westerly force. A chain of spiralling low pressure systems brings great storms followed by eerie days of calm: a manic climate.

In the Southern Ocean, European navigators found that gritty, claimable Earth was fictive, wishful, and disappointingly finite.[11] In this region of delusional geography, ice and mist blurred the boundaries between water and air, and sometimes even posed as land. Explorers in the high southern latitudes had to learn to read the sea and the sky for hopeful signs of land: they scrutinised floating kelp, swarms of birds, a change in the colour of the water, quixotic winds, big black clouds, and large, discoloured icebergs. In the screaming fifties, surging and converging ocean currents generate submarine fronts. Here is found perhaps the longest and most significant biological boundary on earth, the Antarctic convergence, where cold Antarctic waters encounter and slip under the warmer sub-Antarctic waters, and the boundary is manifest in mists and cold and sudden southern marine biodiversity.

The contours and constancy of wind in these latitudes was made visible by a bird. The forties and fifties quickly became known as 'the albatross latitudes' because they were dominated by this magnificent creature that visited and followed ships. Ten of the world's thirteen species of albatross live on the open seas of the Southern Hemisphere. It is particularly the domain of the wandering albatross, pure white with jet black wing tips, the largest surviving flying animal. The albatross is more than a creature of the wind; it was, in the poet Coleridge's words, 'the bird/ That made the breeze to blow'. Voyagers long anticipated its appearance in the albatross latitudes and recorded the sightings in their diaries and journals, for this bird was the signature of the wind that also filled the ship's sails and took them east. Any vessel was attractive to the bird because it created eddies and up-draughts in air and sea; a ship churned the shrimps and squids to the surface, and there was always galley refuse. Albatrosses are surface feeders and swoop at any floating scrap, particularly if it is white. A hundred years ago, Dr Edward Wilson on Robert Falcon Scott's ship, *Discovery*, on examining the stomach of an albatross, found in it 'an undigested Roman Catholic tract with a portrait of Cardinal Vaughan'.[12] Shooting albatrosses from ships was a ritual sport, in spite of the warning in Coleridge's *Rime of the Ancient Mariner*. The bird's light, air-filled bones were claimed by sailors for pipe stems. The ease of flight of the albatross was entrancing. It was so big, like a clumsy goose when lured onto the deck with titbits from crew and passengers, but in the air it soared, hardly ever beating its very narrow, very long wings. Albatrosses inhabit that thin, lively layer of encounter between ocean and moving air, and they would follow ships for days, carried along by the same medium. In low winds and calms they have to forsake the air and rest on the water. For the Wandering Albatross, land is only important for nesting, and a young bird may fly the open seas for seven years before returning to land to breed.[13]

The Great Circle

Australian history was shaped by the European discovery of the sea and the winds. In 1519–1522, Magellan's fleet — in spite of the loss of three ships and 200 men along the way, including the captain — circumnavigated the world and discovered the astonishing unity of the oceans. The world turned out to be larger than Europeans had supposed, and a third great ocean — the Pacific — was found and crossed. In the southern latitudes they found that the sea and the wind flowed with awesome continuity.[14]

If you could master those winds, you could dominate the world. The discovery of the sea introduced an era of maritime empires where expanding trade networks and political influence fell into the hands of those states that could build reliable ships and harness the power of winds. In his book *The Discovery of the Sea*, J. H. Parry attributes early Portuguese maritime success to its location at 'the street corner of Europe' where it bridged the Mediterranean and north Atlantic worlds, and where, for most of the year, the winds off the coast favoured exploration to the south and south-west.[15] Europeans learned not to sail down the west coast of Africa, where they were repelled by northbound currents and winds, but instead to track across the south Atlantic almost to Brazil where they would catch the roaring forties back east.[16] In the seventeenth and eighteenth centuries, the Dutch, French and English became the maritime superpowers who destroyed the older Arab and Venetian trade between the East and Europe by mastering the alternative sea routes. Dutch seafarers rounded the Cape of Good Hope at 40 degrees south and took the prevailing westerlies as far as they could before curving late and northwards towards Java. Hence Australia's west coast was an accidental European landfall, its discovery a passive product of a game with the wind. For the same reason, Australia's more fertile east coast was for long protected from the European imperial eye by the west winds, because heading into them from Cape Horn at high latitude was unhappy sailing.[17]

The great historian of the roaring forties is Geoffrey Blainey who wrote these winds into the fabric of Australian history in his book *The Tyranny of Distance* (1966).[18] His abiding interest in climate and culture, and his fascination with the historical physics and geography of the globe, enabled him to offer new and enduring insights into the foundation of British settlement in the Pacific. Australia was colonised by the British less for its land than for its strategic position in an ocean, and it was indeed as an archipelago of islands that it first developed its European character. Safe harbours, timber for masts, flax for sails, proximity to the chief ports and trade routes of Asia: these were the sinews of the eighteenth-century struggle for sea power.

It is worth recalling the sea consciousness of Australia's early European settlers in the years before their national imaginations turned decisively inland. Australia was invaded by a naval power, its first colonial culture of authority was maritime, whaling and sealing were the colony's earliest productive industries, and it took settlers a quarter of a century to cross the first land barrier, the Blue Mountains.[19] Early Sydney revolved around its port life, its population was 'waterbound' and maintained 'a constant and keen-eyed watch over the harbour', lives were animated and threatened by the cycles of winds, currents, tides, the dangers, risks and opportunities of the sea.[20]

Experiencing the roaring forties was like an initiation ritual for Australia-bound voyagers. What was it like to be on a ship driven by the 'brave west winds'?

'Running before a Gale': Edward Snell's sketch of his ship, the *Bolton*, bound for South Australia at 44° South in November 1849.

SNELL'S DIARY 1849–1859, AUSTRALIAN MANUSCRIPTS COLLECTION, STATE LIBRARY OF VICTORIA

Englishman Edward Snell emigrated to Adelaide in 1849 on board the barque *Bolton*. He had always felt a yearning for the sea, and here he is writing in his diary at latitude 36 degrees south, and heading for the forties:

> A regular unmistakable thorough hard gale of wind ... The ship behaved very well riding over the seas like a duck though they were running as high as our main top. The roaring of the wind was tremendous and the ship rolled and pitched very heavily. After tea I went on deck to get a smoke and was immediately thrown off my legs and struck my right temple against the deck with such a force that I was all but stunned, didn't break the pipe though.[21]
>
> ... Came on to blow pretty hard in the evening and before sail could be taken in the fore top gallant sail went to smithereens, the ship flying through the water 11 knots an hour steering E.S.E. ... The scene was magnificent for it lightened and rained tremendously but the thunder was drowned by the roaring of the wind, for some time it blew a little hurricane, and I thought of the song
>
> 'With lightning above us and darkness below,
> Through the wild waste of waters right onward we go.'
>
> And what a difference there was between singing it in a drawing room at home and looking at the reality at sea. However without any humbug the thing was beyond a joke, and for sometime the sailors had enough to do to get in the canvass which was threshing and flapping about as if it meant to carry the mast away while the rain came down in buckets full and sailors don't use umbrellas.[22]

Snell returned to England nine years later, in 1858, on the clipper *Norfolk*, a vessel so built for speed that Snell believed that 'it would go ahead even in a calm',

The *Norfolk* under snow at 60° South as it takes Edward Snell on the Great Circle route home to England in April 1858.

SNELL'S DIARY 1849–1859, AUSTRALIAN MANUSCRIPTS COLLECTION, STATE LIBRARY OF VICTORIA

although its narrow beam meant that this diarist found it 'rolls like the devil which makes writing rather difficult'.[23] The return journey was by way of Cape Horn, and again the ship dived into the engine of the westerlies, but this time much further south. From Melbourne it dipped immediately south to a latitude of about 60 degrees and maintained this parallel right across the Pacific and beyond Cape Horn:

> ... weather alternately calm, and stormy [records Snell], but always wet, foggy, misty, drizzling, and infernally cold. ... We are today well round Cape Horn but with the decks covered with snow, sails frozen stiff as boards and every one with a drop on the end of his nose, except the ladies who don't go on deck much.[24]

The latitudinal difference between Snell's two voyages — the first in 1849, the second in 1858 — was not just a function of having to round the more southerly Cape Horn, it was also due to the fact that the 1850s had seen a revolution in the sailing of the Australian route. In 1842 an American naval officer, Matthew Fontaine Maury, began to seek the logbooks of captains so that he could compile an atlas of the world's winds, storm systems and ocean currents, and from 1851 he published a series of wind and current charts to assist sailors. At the same time, an English watchmaker, John Towson, advocated the use of the Great Circle sailing route to Australia, because use of his inexpensive chronometers would enable captains to monitor longitude and thus take a short cut around the globe. As Blainey put it, Towson 'believed that the world was round and that seamen should recognise that truth'.[25] The shortest distance between two points on the Earth was the arc of a great circle, and the largest expanse of ocean, the strongest winds, and the greatest distortions of the old Mercator charts were all on the high latitude Australian route.

Captains began arcing down to the furious fifties to save a thousand miles. Matthew Maury enthused that the round trip to Australia affords 'the bravest winds, the fairest sweep, and the fastest running' along a route which for more than 300 degrees longitude courses with the westerlies.[26] 'The billows there lift themselves up in long ridges with deep hollows between them', he wrote; 'their march is stately and their roll majestic', 'the scenery among them is grand'. Maury's passion was to understand the planet's 'two oceans of air and water', as he put it, and to show the practical advantages of philosophical research. He noted with pride that his charts and the Great Circle sailing had reduced the voyage from Europe to California by a month and a half, and the round trip between England and Australia by fifty days or more.[27] His famous book, *The Physical Geography of the Sea*, was first published in 1855 and remained in print for twenty years, its public life corresponding to the triumphant years of the clipper.

The American clipper evolved its finest and fastest form in the same years that gold was discovered in the Australian colonies and the science of wind developed such practical dimensions. Gold, tall ships and the mastery of wind together revolutionised Australia's economic growth, and also secured Melbourne's commercial dominance in the nineteenth century.[28] Melbourne was well situated as the major Australian port for the wind travellers for it combined a rich hinterland with proximity to 'one of the great natural propulsion bands of the earth'.[29] The winds that brought immigrants so swiftly to Melbourne, though, were the very winds that Dr James Barrett thought that itinerant consumptives should be escaping from.

The breath of Antarctica

In December 1789, the supply ship, HMS *Guardian*, with 123 people including twenty-five convicts, and carrying 1003 tons of stores for the infant colony of New South Wales, topped up with 2000 gallons of wine from the Canary Islands, hit a giant iceberg at 40 degrees south. The ship had been painstakingly fitted out in response to Governor Phillip's written cries for help in the difficult first year of the colony's founding.[30] The commander of the *Guardian*, Lieutenant Riou, recorded the sudden vision of 'a body of ice full twice as high as our masthead, showing itself through the thickest fog I ever witnessed ... The ship at this time seemed to be entering a cavern which was large enough to receive her entirely'. The bulk of the cargo was got rid of, some of the men abandoned ship and were never seen again, and over eight agonising weeks the *Guardian* eventually limped back to Cape Town, where salvageable stores were forwarded by available vessels to Sydney. The loss of the *Guardian* was a bitter blow to the struggling British settlement, one that Governor Phillip judged to be 'almost fatal to the Colony'.[31] The ice of Antarctica, roaming wild in the roaring forties, had already intervened dramatically in Australian history.[32]

The icebergs were mysterious, far-flung evidence of what lay at the south pole. What was their source? Where did the roaring forties come from, and go to? Just as the classical cartographers intuited Antarctica's presence from arguments of earthly symmetry and elemental equilibrium, so did early voyagers sense the ice continent before they saw it: they felt its breath. James Cook circled Antarctica in a dance of awe. He voyaged distastefully through seas 'pestered with ice' and amongst icebergs that left his 'mind filled with horror'; he made what was perhaps his greatest discovery, the circumpolarity of the Southern Ocean.[33] Cook's three summers probing the Antarctic Circle, in 1772–1775 in almost constant bad weather, came near the end of the Little Ice Age, an era of lower global temperatures and more extensive pack ice, and this may explain why this ambitious navigator did not conclusively sight the continent.[34] And when Cook passed close to Heard Island in February 1773 but did not see it, the towering feature may have been disguised not only by cloud, but also by a colder climate that clad it in ice to the sea.[35] By the time of Great Circle sailing eighty years later, the Little Ice Age was ending, and high latitude voyaging may have just become, with fortuitous timing, a little less perilous. For much of the nineteenth century, sailors and scientists analysed the sea and the ice and the wind as indirect evidence of the character of Antarctica, and in the early twentieth century, the effect of the southern distribution of ice on Australian weather was one of the key research topics suggested for the officers of the new Bureau of Meteorology.[36]

Scientists began to realise that understanding the Southern Ocean and the influence of Antarctica would offer clues to Australia's climate. Henry Chamberlain Russell, government astronomer and meteorologist in New South Wales, had already provoked debate with his finding in 1893 that in the Southern Hemisphere, anticyclones were not fixed but moving, and that they marched across central Australia in a relentless and predictable progression.[37] He explained this moving chain of pressure systems by the great predominance of ocean over land in these latitudes. A similar progression of cyclones tracked across the Southern Ocean, and while the northern edges of these low pressure systems affected Australia, their southern edges extended over ice, and their centres lay unmapped over the Southern Ocean of the roaring forties and screaming fifties.[38] Writing in 1859, Matthew Maury hoped that one day meteorological stations would be established on Heard Island and other points in the region of the 'brave west winds'. 'I know of no enterprise in the meteorological way that ... gives promise of richer rewards than this does, both practically to the mariner and scientifically to the philosopher'.[39]

The study of cyclones was revolutionised in the early twentieth century by the Bergen School of meteorologists led by the Norwegian, Vilhelm Bjerknes. He brought a physicist's eye to atmospheric science — seeing air as a fluid — and he pioneered a more sophisticated analysis of air masses and fronts. In the years following the Great War, strategic forecasting became more urgent and air travel common, and improved communications enabled a dense network of weather observations to be amassed quickly so that finer-grained atmospheric structures could be promptly mapped. The frontal concept of the Bergen School introduced three-dimensionality and dynamism to the study of air masses, and cyclones were seen as evolving phenomena with life-cycles that might be studied. Bjerknes also outlined his idea of a polar front, which he saw as an oscillating battleline between polar and equatorial air, an undulating string upon which the pearls of cyclones developed and moved.[40] Australian scientists did not systematically test these concepts until the Second World War and after, but meteorology was at the forefront of early twentieth century Australian investigations of the Southern Ocean and the polar continent. The links between the Gondwanan partners were sketched out in the atmosphere, as well as in the rocks and fossils.[41]

Leaning on the wind: southern meteorology

No continent is more ruled by wind than Antarctica. Mawson famously called it *The Home of the Blizzard*. He established his base at Commonwealth Bay on the Antarctic continent in 1912 on a site that he later found to be probably the windiest place on earth. The average wind there is an extraordinarily steady gale.

At Commonwealth Bay, the meteorologist C. T. Madigan left memorable descriptions of his work recording a continental weather pattern ruled by wind: not the west winds, but the violent gravity-driven or katabatic winds that dominate Antarctica itself. They are surface winds formed of cold air pouring down off the high Antarctic plateau; they can become air avalanches. In 1912, complete weather observations were taken every six hours — these consisted of reading the barometer and screen thermometer, estimating the wind velocity and noting its direction on the vane, describing the cloud, recording any phenomena like snowfall, drift, St Elmo's Fire, coronas and halos, sunset colours, state of the sea, and auroral displays. All data was entered in pencil in a meteorological day book, and fifteen of these hundred-page books were filled. Visiting the anemometer was, admitted Madigan, 'physically a fairly severe undertaking, presenting considerable difficulty in the mere finding of it'. He goes on to explain:

'Leaning on the wind': a member of Mawson's expedition at Commonwealth Bay, Antarctica, quarries ice for cooking purposes.

FRANK HURLEY, NATIONAL LIBRARY OF AUSTRALIA

> After some practice the members of the expedition were able to abandon crawling, and walked on their feet in these 90 mile torrents of air, 'leaning on the wind'. The picture over this title ... shows a man using a pick in getting ice for cooking purposes, leaning comfortably on the wind as against a wall. For the sceptical one might point out that the body is at rest and in sharp focus, while the pick is moving and somewhat blurred. I found that in a 90-mile wind, if I faced it and kept the body straight, I could touch the ground in front by extending an arm. Of course, one wore more clothing and offered a larger area to the wind than under normal conditions.[42]

Mastery of this 'hurricane walking', as they called it, became an essential meteorological accessory. Douglas Mawson declared that: 'What is certain in my mind is that like difficulties have never yet been experienced in the same degree at any other meteorological station. The bare records in their unfailing regularity are an epic in meteorological achievement'.[43]

Meanwhile at Queen Mary Land, Morton Moyes monitored the weather for the Western Party led by Frank Wild. On rough days, Moyes had a similar battle to collect the daily observations:

> We got back against the wind with difficulty, and I had to stand inside for half an hour with arms up, as hands were frostbitten. Wind so strong, unable to stand up against it, and could not see [my] hand in front because of snow.[44]

Amidst grasses flattened by the westerly winds, George Ainsworth leans against the stevenson screen at Macquarie Island where he was meteorologist from 1911–1913.

For ten weeks Moyes was left completely alone with his weather instruments at base camp, feeling 'like the last leaf of a branch'. Between reading the wind gauge and the snow gauge, he did some other reading: Shackleton's *The Heart of Antarctica*, the meteorological notes of the previous British and Scottish expeditions, Macaulay's *History of England*, Browning's poetry, Spinoza's *Ethics*, Dante's *Inferno*, Harmsworth's *History of the World*, Bacon's Essays, Geikie's *Geology*. He wrote in his diary: 'I'd like a good novel for a change, the shops here don't stock them'.[45]

A priority of the Mawson expedition was to establish a weather station at Macquarie Island, thereby fulfilling Matthew Maury's dream of doing meteorology in the teeth of the westerlies. It was established with the strong support of Henry Hunt, the Commonwealth Meteorologist, who loaned one of his officers, George Ainsworth, to lead the island party.[46] A wireless mast was erected and daily weather readings were communicated to Australia and New Zealand. An earthquake tremor greeted the party on arrival in December 1911, and the wind 'appeared like an invisible wall'; you had to turn your back on it to take a breath.[47] One of the meteorologists sent down later to Macquarie Island took a football with him but reported that, because of the wind, it 'could not be used'. The original party built their shack on the easterly side of a huge rock that protected them from the relentless westerlies, and at times they had to catch practically all the food they ate. They lived off the fat of the land — the heart, tongue and liver of the locals. Ainsworth confessed to Hunt that 'I shall never willingly look a Sea elephant in the face again'.[48]

In the mid-nineteenth century, Matthew Maury had speculated that there was a mild climate about the south pole, mild by comparison with the Arctic. 'The winds were the first to whisper of this strange state of things', he wrote, and the other clues he drew on were those ships' logs that he knew so well which recorded low barometric pressure at high southern latitudes. He extrapolated their descending oceanic readings into the expectation of a permanent, very low pressure system at the south pole, whose latent heat and water vapour would suck in moderating winds from the north. He thought that the roaring forties — the winds that drove his sailing ships — must go all the way to the pole.[49]

The first human did not step foot on Antarctica until the end of the nineteenth century. It took even longer for people to realise that there was not just land down there but a continent, that it was high and dry, and that it was very, very cold, much colder than the Arctic.[50] Its meteorology, they were to discover, was distinct, extreme and simple, all of which made its effect on the rest of the world's weather greater. The south pole harbours a permanent high pressure system and easterly winds blow round the continent's coast. Those outer westerlies of the forties and fifties, acting in concert with the circumpolar ocean current, isolate and amplify Antarctica. They form an atmospheric equivalent to the ocean convergence, and both boundaries — the polar front and the convergence — were not recognised until well into the twentieth century.

Antarctica is 'the world's greatest refrigerator', as Mawson called it in 1935;[51] it is such a cold core of the atmosphere that it 'deflects the meteorological equator of the globe northward nearly ten degrees latitude'.[52] The continent has 90 per cent of the Earth's land ice and locks up 70 per cent of its fresh water; all that permanent ice reflects the sun in summer, and the dry clear atmosphere releases what heat there is. Unlike the Arctic, the solar radiation balance of Antarctica is negative all year round. 'The ice amplifies the conditions that generate the ice' by the same feedback mechanisms that, on a global scale, can produce Ice Ages.[53] Over the last century, we have come to realise that Antarctica is not only a powerful influence on Australian and world weather, but it is an essential archive of past climates, and a crucial indicator of global changes today. The roaring forties are both a product and an agent of the ice. And the ice holds secrets of the history of the air.

Tom Griffiths

Tom Griffiths is an environmental and cultural historian whose books include *Hunters and Collectors: The Antiquarian Imagination in Australia* (Cambridge University Press, 1996, winner of the NSW Premier's Book of the Year), *Ecology and Empire: Environmental History of Settler Societies* (co-edited with Libby Robin, Keele University Press, 1997), and *Forests of Ash: An Environmental History* (Cambridge, 2001). He has collaborated with the artist Mandy Martin in an evocation of the Channel Country of south-west Queensland (Mandy Martin *Inflows: The Channel Country*, 2001) and with Tim Bonyhady in editing a collection entitled *Words for Country: Landscape and Language in Australia* (University of New South Wales Press, 2002). He is a Senior Fellow in the History Program of the Research School of Social Sciences at the Australian National University, and a Fellow of the Australian Academy of the Humanities.

14

Cyclone Tracy
voices on the wind

Bill Bunbury

Dramatis Personae (in order of appearance)[1]

Sandra Holmes	*Darwin resident*
Echo Coles	*Aboriginal leader*
'Strider' (William Walsh)	*Relief worker from Darwin*
Earnest Young Journalist	*From a southern city*
Barbara James	*Darwin resident and historian*
Ella Stack	*Medical practitioner*
Dawn Lawrie	*Darwin resident and Northern Territory member for Nightcliff*
Peter Dermoudy	*Darwin architect*
Beth Harvey	*Mother of three, evacuated from Darwin*
June Dunstan	*In charge of the Red Cross processing point at Perth airport*
Erica Wiltshire	*Former Darwin resident*
Howard Truran	*Former Darwin resident*
Jack Haritos	*Darwin resident*
Anne Ross	*Former Darwin resident*

Setting

Tropical cyclones are a dramatic feature of the distinct coastal climate of Australia's north. Rainfall on the coastal plain is high and reliable, but also highly seasonal — for this is monsoonal Australia, a land governed by the annual cycle of wet and dry. From December to April each year, 20,000 kilometres of the northern Australian coastline are vulnerable to cyclones. They develop in the monsoon trough or in a deep moist easterly flow over warm ocean water to the north. Cyclones are 'potentially the most dangerous of all meteorological phenomena'.[2]

Prologue

There were three phases to Cyclone Tracy, which struck Darwin with such ferocity on Christmas Eve, 1974. It was the worst storm in Australia's recorded history, destroying almost an entire city and taking at the very least sixty-five lives, probably more, as there were almost certainly unrecorded deaths among the itinerant and temporary population of Darwin.

Destruction and dispersal reflected the elemental power of the event and human fragility in the face of such odds. Renewal followed, but it came at a cost.

Cyclone Tracy is a social metaphor for both dispersal and renewal. A cyclone reaches maximum velocity in one direction — there is a temporary calm, the eye — then blows with equal power in the reverse direction. The events that followed 24 December 1974 parallel that cyclic movement to some extent. Tracy blew away many human relationships but strengthened others. It tore the social fabric. For a time, Darwin was a one generation town, without children and grandparents. Outsiders came in to rebuild and police the city. Many who stayed felt they had little control of their lives or their city.

Tracy was a harsh reminder of the notion of 'frontier'. Darwin sits at Australia's northern edge and lies exposed to tropical turbulence and, as we've seen in more recent times, political turbulence with events in East Timor, Bali and other parts of Indonesia.

Tracy also sharpened a sense of what was (and was not) home. Losing property and possessions fostered a consciousness of heritage. So many Darwins had already been destroyed by previous cyclones and wounded by war that the city's reconstruction after Tracy posed questions of recognition and familiarity. The new 'Canberra in the tropics' that emerged was not the Darwin older Territorians remembered.

These are some of the themes I dealt with in the radio series I made in 1987, thirteen years after Tracy, and later in a book written to commemorate the twentieth anniversary of Tracy, *Cyclone Tracy: Picking up the Pieces*.[3]

Act I: Destruction

The notion of elemental power has suffused our literature and our history since we were first able to record it. Homer's map of the long and perilous journey of Ulysses from Troy to Ithaca is not only an evocation of the sometimes playful, sometimes vengeful power of the gods but it also vividly depicts a capricious, almost land-locked ocean — the Mediterranean — given to sudden and unpredictable tempest and destruction. In our largely urban world it is easy to lose touch with this sense of the power of nature. But some Darwin residents in 1974 were still well attuned to it:

Sandra: On the day it was just dark, gloomy and sweltering hot and the first couple of frogs started to croak. As the day wore on there was a frenzy of sound, which was almost deafening.

Echo: The cyclone hit us Christmas morning, but about 4 o'clock Christmas Eve I noticed a lot of our people were just moving down the track.

Sandra: There were strange things happening. Feverish activity of ants who were madly cleaning out their nests and doing all sorts of things. Then with the first rains beetles were falling from the sky, thousands of beetles. Things were rushing against the flywire, you know, in a mad frenzy of insect life and it made one uneasy.

'Strider': That particular morning I woke up early and I was far up in the hills somewhere, sort of a block where I could see for miles and miles and miles. In the early morning just as the light came there was a column of mist that went up every piece of water in the whole district. But even very very small water holes and rock holes and things like that were marked on that morning by this column of condensation rising vertically straight up in the air. I'd never seen anything like it before in my life. I remember my grandmother always telling me about like a bit of windstorm or sandstorm coming through but I did have that feeling that we were going to get hit by Cyclone Tracy.

Sandra: Then some of the old white blokes said, 'Mmm, you know, look the mangoes are falling off the tree, mmm, that's bad'. And we had two earth tremors, you know. The Aborigines saw the feverish activities I did.

Echo: Everything appeared dead, there was no bird life, no movement. Even the trees were still.

Aside

Earnest Young Journalist (*impressed by the perceptiveness of Darwin's Indigenous people, who had walked away from Darwin well ahead of the cyclone and hoping for a revelation of their sense of the numinous*): How did you know you had to leave?

Cyclone Tracy's destruction.

IMAGE COURTESY OF JACK HARITOS AND FREMANTLE ARTS CENTRE PRESS

Aboriginal leader: Oh! We just watched the ABC TV weather forecast.

The radio series began with a program, Making Deals with God All Night, *a quote from a survivor, reflecting the sheer helplessness in the face of such an unexpected and terrifying natural force. The most immediate fear, described by several witnesses, was that this disaster had only happened to them, or that they were the only survivors.*

Barbara: I got to the back door, opened the door and the cat just flew out of the house. I had to stop and wait for the bedroom walls to go away and the bedroom furniture to fly past before we could make our way downstairs. The only thing we could see was a fireplace outside, which wasn't very practical so we went to the garden. The first thing that happened to me was a loose power line wrapped itself around me like a whip. That was probably the moment of greatest panic because I really felt the wires were live. I remember at that moment thinking that we could die.

Sandra: Then the house started to shake and rock, then it was just survival. We had a big heavy mahogany table — we had glass doors at one end of the house but little panes of glass — and we kept hurling this back at the doors, and all around you were spears of wood flying like shrapnel.

The houses that collapsed all around me, all their spears of timber were coming through as fast as bullets. You were hanging on and you were hurling this table again and again to try and hold these doors and then you were hurled back. The wind gusts were terrifying.

Then you would rush forward again with the table, and this went on all night. It got terribly cold, icy cold and you were shouting at each other. I was praying all the time like a litany that never stopped. Asking for our salvation — I guess — from God.

Ella: Most of Darwin made deals with God all night. I changed mine as the night got worse. First of all I informed God He had to save the whole family and then as things got worse I rather more humbly made a deal that my husband and I could go because we'd lived a good life but the children ought not to go.

Dawn: I kept thinking all night — what have my children done to deserve this? Because I was quite sure we were all going to be killed. That was all I could think of, what have they done? Why them?

Ella: I said to one fellow who was firmly agnostic, 'You must have called on some power outside yourself because there was no power in yourself to do anything'. He said, 'Well I eventually said, "Well God if you are there, if there is a God will you help my family?"' But during the night it was noise, it was fear and it was helplessness because you knew there was nothing you could do. No matter how well-trained, how clever you are, how well your brain works, you could do nothing.

Not long before the cyclone, Peter Dermoudy had been on public record warning that some newer buildings would not withstand cyclonic pressure. As Tracy grew in intensity he sought shelter at the safest place he could think of, the East Point concrete Gun emplacement built to repel the Japanese in World War Two. From its slit window he watched the progress of the cyclone for much of the night. But he feared that he was the only survivor.

Peter: It was quite strange because there had been a whole number of films and books and things on this particular subject. You know particularly after the atomic stuff. I was beginning to wonder, I was starting to really plan, I was starting to say, oh well there must be underground fuel so I can always get a pump of some sort to keep my car going. Then I thought there must be some transceivers so I can get some messages out to outside. There must be something left somewhere even though all the people have gone.

Then I drove down past the Fanny Bay Hotel and that was a wreck. There was nobody there. Drove for about another half mile and all the houses had gone. I think it's that experience that a lot of people had where they thought they may have been the only people left alive.

Ella: I thought there would be nobody alive in the morning, including us.

Sandra: Then gradually the buffeting eased and a grey light broke and I looked around and I said to my husband, 'We must be about the most unlucky people in Darwin'. Then looking through the broken doors I saw and heard the distant sound of ambulance sirens and a red light flickering and I said, 'Somebody must have been hurt'. Which of course was an understatement.

Then looking outside, all around the houses were gone. The power poles were twisted and all the wires hanging like crazy spiderwebs. Opposite, our neighbour, Carl, with his children, his house had gone completely, only the stumps left. But they crawled out of the rubble and he actually found a bottle of Scotch. Christmas has never meant quite the same to me since then.

Act II: Dispersal

Major General Alan Stretton, who headed the Natural Disasters Relief Operation and worked out the evacuation plan, brought in the next phase — dispersal. In Stretton's view there was insufficient food in the city to sustain its population. Power supplies had gone and with them the sewerage system and ability to preserve food. Disease was a definite risk. Darwin must shed its population if it was to avert a major health crisis.

In the airlift that followed, 30,000 people described as non-essential were sent south to friends or relatives, leaving 10,000 to stay and rebuild Darwin. Inevitably this meant splitting families. Older people, those who were frail and mothers with young children went out.

Beth (***at the airport***): Well, we just stood. We registered then with the people behind the counter. I don't know who they were but we registered and said who we were and where we were going. The luggage labels were written out with our names and destinations on and attached to the children like they were pieces of suitcase. There were no tickets issued or anything like that. The toilets were absolutely dreadful because there was no water and so there were buckets of water but there was no light in the airport, no air conditioning, it was steamy, there were crowds. There was nowhere to sit, people were standing with their suitcases and then as a plane would go the next lot of people would go out on to the tarmac from inside the building. Then the next lot of buses would be unloaded.

Jack Haritos surveys the ruins of his house in Darwin.

IMAGE COURTESY OF JACK HARITOS AND FREMANTLE ARTS CENTRE PRESS

As we went out the door they were still checking and checking where people were going. As we got to the door, that was the first time I ever panicked in the whole situation because Paul and Wendy — Ben wasn't walking, I was carrying him. I was carrying a bag, oh carrying a suitcase and I had an overnight bag as well and carrying Ben too and Paul and Wendy were in front of me and two women pushed between me and the two children. I totally panicked and the man's saying, 'It's all right, keep your shirt on, it will be all right'. I said, 'They're not your children there' ... I think that was the first and only time I panicked the whole time.

Not all were reluctant to leave:

June: One of my volunteer staff [*receiving Darwin refugees at the Perth airport*] came to me for help, saying that she was trying to deal with a middle-aged woman who would not give any personal details. I took the woman aside. I explained that it was standard evacuation procedure to let relatives know she was safely down and where she would be staying, and why it was necessary to let her husband or family know that she was in Perth. 'Are we alone?' she asked me confidentially. 'Yes', I reassured her. 'I've been trying to get away from the old bugger for years and this has been my best chance yet!'

Long separations tested and changed relationships — relationship to the place Darwin people thought of as home, and also partnerships and marriages. Tracy also tested people right from the start — simply in coping with a major disaster.

'Strider': The other general conclusion or speculation that I would like to make about the cyclone, something that intrigued me, was the role that was played after the cyclone by what I call 'hippie people'. I mean I didn't go everywhere but certainly in the parts of the town that I moved in they were very prominent. They got organised very quickly for people who traditionally are not organised and don't like organisation. I guess people who are accustomed to a fairly high level of confusion and uncertainty have got a high tolerance for uncertainty. They performed very well after the cyclone, they didn't seem to be too disturbed by the fact that everything was different, and they were very practically minded. I guess everyone was practically minded after the cyclone but they were a bit quicker to actually get things organised, like find some way to cook and so on. By the end of that first day there were mainly hippie people in various emergency kitchens and things around the place so they very quickly got themselves organised to do what had to be done. I thought that after a few days when the shock wore off people resented that and were not pleased when they discovered that they had been dependent on hippies for the previous few days.

Inevitably, as Darwin began to recover physically, those separated from their partners wanted to return. But the authorities slowed the process. In their view the city had to become properly habitable before families could move back. Official policy conflicted sharply with the human desire to re-establish home. Those who tried to come back had to cope with the permit system, designed to ensure that people could come back to proper housing, but which conflicted with the natural desire of people to rebuild.

Dawn: The authorities introduced an iniquitous permit system which I believe was unlawful. Where a Darwin citizen wishing to return to their home to rebuild their lives and make their future, rebuild their home, was told unless they had a permit they couldn't come back. Now these permits were issued by petty bureaucrats who would ask questions as to, 'And where are you going to live?' At this stage I was involved in that system, I was handing out addresses of houses that were habitable so people could say, 'Oh I'm staying at 26 Smith Street', or whatever. Until they woke up that all of a sudden there were hundreds of people giving the one address. But we were deliberately finding ways around the system. It was dreadful to say to people you can't come back.

Even worse if a woman who had been evacuated and was applying to come back to Darwin and her husband was here. They would ask her husband if he wanted her to come back.

The permit system was abandoned in June 1975 in the face of widespread non-compliance. Cyclone Tracy also blew away the town's children. Dawn had been one of the first to send her children south and one of the first to bring them back.

Dawn: Other people hadn't been so fortunate. Their children were still south, we missed them and they missed us. Now when we brought them back, I'd walk down the street and they'd come over and pat my kids on the head. They were hungry for children. As I've said in another place I realised what a dreadful thing the Pied Piper did to Hamelin when he took the children because the society ceases to exist. I think I was a catalyst for people insisting they would bring their children back.

Act III: Renewal

The final element in Tracy's cycle was renewal. It came at a cost. The physical and financial costs of the cyclone were all too obvious. Less conspicuous, but deeply felt, were the personal effects on individuals of all ages of their experiences both in the storm and in the aftermath.

Beth: When we went to Melbourne initially, Wendy asked my aunt: was her house blown down? She couldn't comprehend the distance between Darwin and Melbourne because we'd sort of arrived there in no time flat. She thought everybody's house was damaged.

There was a bit of a wind one day when the children were in the Melbourne backyard. The door was banged and the wind you know, the trees flapped and what have you. Paul and Wendy were about to run in and Angela turned round and said to them, 'Don't worry about the wind here. It's only in Darwin it blows down houses, it doesn't here'. And Paul said, 'Doesn't it?' And Angela said, 'No, just Darwin'.

And they really accepted that. Whereas I was ready to race out and comfort them, which would have been the worst thing in the world to do because it would have reinforced their fear of the wind. But we just sat back and laughed. I was glad I didn't race out at that stage because I wouldn't have been much help. But they still were edgy with the wind initially.

Erica: I found for quite a long period after that my memory very suddenly deteriorated very badly. Names of people, names of books, names of authors, names of historical ... you know all sorts of names eluded me and that didn't ever entirely repair but gradually by working hard at it, it has improved. But I think that was one of the quite unimportant but rather distressing forms of shock. I was amongst the extremely lucky ones. For a lot of people the situation on the night of the cyclone was much more traumatic. I mean we were four together, we could see that we were all right even if it was very unpleasant and frightening.

But others whose husbands were called on duty and whose houses went to pieces around them, I think, were absolutely shattered. Others lost family members, while others had family members who were grievously injured, and were straightaway evacuated. People suffered in an enormous variety of ways.

'Strider': Yes I wasn't physically injured but I am sure that the psychological impact was enormous. I went to Melbourne to do a job straight after the cyclone and it turned out that I couldn't do it. Looking back on it now I just think it was laughable to try and do any sort of major serious job a couple of weeks after a cyclone. Just ridiculous to ask somebody to do something like that. I reckon that it took me a few years to get over the major mental shock of the thing.

There was also physical and spiritual weariness. Howard Truran had lived in Darwin for a long time before Tracy and had fought the authorities to bring his family back. Now he was reconsidering.

Howard: I would say that between the time I started work after the cyclone and the family was back, and I was coming home in the bus every night and I was getting off at the stop and I was walking past all the ruins and that, I must admit that I did become a bit depressed and a bit tense. Then at the stage where we were looking to move and leave Darwin — that surprised a lot of people because I'd been up there for a while. So they just said, 'We never thought you'd ever leave Darwin'.

Possibly — we've talked about it since — if it hadn't been for Cyclone Tracy we would probably have never left Darwin. I'd had enough. I'd worked hard on the house and we more or less pioneered the place from scrub bush when we were felling trees, when we first moved into the house. Felling trees and all that sort of stuff. And I saw all the garden and everything wiped out and I just wanted to get out. I'd become very depressed. Then getting a loan and rebuilding the house and then there was these rumours about they weren't going to let people. The surge line came halfway through our yard.

The federal government knew this when they built these government houses. But it's a line on the map where if a cyclone hit at full tide that's where the high water, you know, 20 or 30 feet of water comes in, and we would have been nearly all under water. So they were drawing maps and they had books out saying, 'Oh well you're not going to be building here', and I thought well we'll have to move elsewhere and so forth. I just couldn't face it, I'd just had enough.

Jack: Oh yes, yes. A lot of people just couldn't face it all again after years of hard work establishing their business, they just couldn't face it.

Erica: I think there were as many varieties of the way it affected people as people. It was undeniably a strain. Not knowing how the husbands were getting on in Darwin, the husbands not knowing, in many cases, how their families were getting on down south. Of course those who could go to friends and relatives, close friends and relatives, were really lucky. Although the relief organisations were working very, very hard, a lot must have been utterly miserable.

Jack: Husbands were separated from wives, husbands came back to see what they could do. A lot of them took to drink. My son was in a gang that went around a few days after the cyclone cleaning up rotten food out of refrigerators. People had left Darwin. He said there were cases where people were sitting round drinking beer and the food was still rotting in their refrigerators. He said he struck quite a few of these homes. There you see people weren't helping themselves. Most people did but these were probably isolated cases. But wives went south, husbands stayed up here. A lot took to drink. Families did break up, yes. A lot of them just couldn't face the prospect of having to rebuild and it broke up a lot of marriages.

Erica: Well for years afterwards we kept coming across people whose marriages had broken up. Tracy might have been the last straw, Tracy might have been an initial trauma and everything in between.

Another interesting perspective in recovery and renewal was the loss of both individual and collective memory.

Anne: A lot of the older people I know were evacuated and because of their age, because of financial situation, a lot of them never did come back. To me they were the story tellers of the early days in Darwin. I think we missed out, you know, we've probably lost a lot of little titbits about Darwin, will never be recalled.

That story telling dimension — the collective memory — reminds us all too well of an often forgotten element in assessing a catastrophic event like Tracy, the essential nature of the culture and the heritage of a community, including its built heritage. The very process of re-thinking Darwin's physical future created a new awareness of its past.

Barbara: What happened with some people was a desire to at least retain the few things that were left and learn more about the history of the place. There were strong moves to save things like the old Town Hall, which had been built in 1883 and the old Court House in 1884. There was such a strong resident action group feeling that most of those buildings were in fact restored. I think that sort of sparked off great interest in what Darwin was and people wanted to know more about it. I think it's in a funny way been responsible for a resurgence of interest in the history of the Territory by the people who live here.

Eleven years on from Tracy, Cyclone Gretel struck on 12–14 April 1985, bringing back the dread and the personal experience of those who had been through Tracy and survived.

'Strider': With that Cyclone Gretel it was interesting at one stage, the wind went round to the south which indicated to me that the cyclone was to my east, so the wind went round for a while. It also started to roar, it started to sound a bit like a distant jet engine and I'd forgotten about that noise, that roaring noise. As soon as it started to roar, it all came back to me with appalling clarity, and I really didn't like it at all, it scared me really badly. I had a really bad fright when I heard that wind.

Epilogue

The cyclonic metaphor has come full circle. Personal memories of Tracy linger. Many of its survivors cannot forget the unforgettable. But Strider also spoke about feeling alienated politically from his own city. Urban planners came from Canberra to create 'Canberra in the tropics'. They met with considerable hostility from locals who wanted to keep familiar streets and landmarks. They reached a compromise but Darwin's reconstruction was delayed in the process. Tracy became part of Darwin and Australia's collective memory.[4]

Bill Bunbury

Bill Bunbury is the producer of the Australian Broadcasting Corporation's Radio National *Hindsight* program. He is the author of nine books on Australian social history including *Cyclone Tracy: Picking Up the Pieces* (Fremantle Arts Centre Press, 1994). In this chapter he returns to this and some of the other themes he dealt with in the radio series for Radio National's *Talking History* program, made in 1987, thirteen years after Tracy. He has won four national and international awards for his radio history features, including the New York Gold Medal in 1996 for best history documentary for *Timber for Gold*, and the NSW Premier's Inaugural Award in 1997 for his six-part radio series *UNFINISHED BUSINESS — Reconciliation, the Republic and the Constitution — Viewed through History's Long Lens*.

Mulvaney Trowel

Hand trowels are the stock tools of archaeologists. This one belonged to leading archaeologist, Emeritus Professor John Mulvaney, who used it at all the sites he has excavated in Australia.

Trained at Melbourne and Cambridge universities, Mulvaney was a pioneer of field archaeology in Australia. He was the first to find proof that Aboriginal people had lived on the Australian continent for many thousands of years, long before European colonisation. In 1962, at Kenniff Cave in southern Queensland, Mulvaney uncovered evidence of human habitation during the last Ice Age, more than 12,000 years ago. Later, this date was overtaken by others of 40,000 years and more.

The work of Mulvaney and those that followed in his footsteps has re-written our understanding of Australia's history. Using radiocarbon dating and other methods, they have given us a 'deep time' perspective of the human occupation of the continent. We now know that people have lived in Australia through remarkable changes in climate, vegetation and animal life. This is an ancient land with a cross-cultural history that spans tens of thousands of years.

Archaeologist's field trowel, used by John Mulvaney, 1956–1975

This well-worn trowel was John Mulvaney's field trowel, used at every archaeological site he excavated in Australia, including Glen Aire, Fromm's Landing, Kenniff Cave and Ingaladdi

NATIONAL MUSEUM OF AUSTRALIA
PHOTOGRAPH BY GEORGE SERRAS

A Change is Expected

Fromm's Landing Latex Peel

Pioneering archaeologist John Mulvaney made this latex peel at Fromm's Landing on the Murray River, in South Australia. Fromm's Landing was the first site that Mulvaney excavated when he began work in Australia, after studies at Cambridge University. In the past, the area was a major centre of Aboriginal habitation.

Archaeologists use latex peels to record the different layers of earth in an excavation pit. They first paint latex onto the wall of the pit, then peel it away once it has dried. Each soil layer or strata recorded on the peel corresponds to a different point in time. Study of the strata in an excavation pit or trench can reveal what climatic events or other conditions occurred in the area in past years.

Mulvaney made this latex peel during work at Fromm's Landing in 1963. It shows there was a large flood in the area about 3,000 years ago. Excavation of sites at Fromm's Landing has also uncovered large amounts of fresh-water mussel shells, bone and hearth material that indicate human habitation stretching back thousands of years. Important discoveries of rock engravings, wrapped burials and early evidence of dingoes have also been made.

Detail of a latex peel from a 1963 archaeological excavation at Fromm's Landing, South Australia

NATIONAL MUSEUM OF AUSTRALIA
PHOTOGRAPH BY GEORGE SERRAS

15

Palaeoclimates
an archaeology of climate change

M. A. Smith

No discussion of climate and culture is complete without at least a glance at the remote past — at the palaeoclimates that expanded or reduced the opportunities for human settlement in Australia. This is a quick sketch of ideas about palaeoclimates from the perspective of an archaeologist. My aim is to remind us that behind the quick flash of weather lies the deep time of changing Quaternary climates, and that we need not just a sense of place, but also a sense of place in time.

The imprint of palaeoclimate

A few years ago, I took my 12-year-old daughter to see Lake Mungo. We talked all morning about ancient lakes and Aboriginal campsites, but looking at the dry scrub and saltbush country, she could not make the imaginative jump. Standing on the lunette, her eyes picked out a tiny piece of bone amongst the drift sand. Less than 10 millimetres long, it was a fish otolith, part of the bony structure of the inner ear of a fish, its shape characteristic of golden perch. Puzzled, she looked around at the dry plain and started to ask, 'How did a fish get way out here?' Watching her eyes I saw the flash of understanding: an ancient lake full of water snapped into focus.

It is fossil landscapes like Lake Mungo — abandoned by the conditions that shaped them — that give some of the strongest impressions of changes in climate. The filling and drying of Mungo was tied to the shifting climate of the last Ice Age. However, not all landscape change is so directly tied to global climate. Landscapes evolve: rivers choke with debris and cut new channels, lakes silt up, hillslopes are destabilised by chance events. And some changes reflect the impact of people.

Disentangling the effects of climate, people, and other landscape dynamics preoccupies the attention of archaeologists and earth scientists. Quaternarists talk of *palaeoclimates*, and use terms such as *proxies* and *transfer functions*, to instil some discipline into how we relate changing environments to changing climate:[1] Which changes can be regarded as a proxy record of climate change? How sensitive is the local environment to climate change? How much climate change is required to transform the bed of Lake Mungo from a world of waterfowl, perch and crayfish to one of saltbush, red kangaroo and emu? And there is always a spatial dimension to these arguments: What parts of the landscape will be affected by a change in climate? At Mungo, lake levels may change — but is the vegetation of the back country affected?

From weather to palaeoclimate

There are other sorts of intellectual transfers to be made as we move from weather to climate to palaeoclimate. *Weather* has an immediacy. It is something experienced, something of the moment. The *Oxford English Dictionary* defines it as 'the condition of the atmosphere (at a given place and time) with respect to heat or cold, quantity of sunshine, presence or absence of rain'. *Climate*, on the other hand, is weather generalised, an abstraction, a perceived pattern or expectation of weather, a system of weather. Climate is the 'condition (of a region or country) in relation to prevailing atmospheric phenomena ... and their effect on human, animal, or vegetable life'. Climate therefore links weather, time, place and land, conditions and consequences, past, present and future into one beguilingly simple concept.

If climate is systemic weather and its consequences for land and people, palaeoclimate is climate writ large, with an emphasis on prior states of the system. Palaeoclimates are those where prediction is retrospective: the weather is reconstructed long after the event and only indirectly from its accrued impact on plant and animal communities, landforms, and people.

Ice Ages

Part of the fascination of studying palaeoclimates is the feeling of being able to peer into a lost world, and this is particularly the case for the last glacial cycle. When I was an undergraduate student, we still worked with the notion of four major ice-ages: the old Penck and Brückner scheme of Gunz, Mindel, Riss, and Wurm.[2] Glacials (cold periods) were separated by inter-glacials (warm periods). These were based on the European alpine sequence of glacial moraines, correlated with faunal and pollen records from other sedimentary sequences. These became the principal building blocks of Quaternary stratigraphy and created a wonderful vision of primordial and primeval worlds to be explored, of — in the words of John Frere in 1800 — 'a very remote period indeed; even beyond that of the present world'.[3] In the 1970s this was already an antique schema. Chicago scientist Cesare Emiliani had explored the use of deep-sea cores as an alternative record of global temperatures, but the full impact of this was yet to be realised.[4]

An orbital metronome

I left Canberra for nearly a decade as a field archaeologist in the Northern Territory. When I picked up the trail again as a Research Fellow at the Australian National University, the Gunz, Mindel, Riss, and Wurm model had disappeared. And with them the idea of *stadial* changes in palaeoclimate. Colleagues in the earth sciences now spoke of Stage 3 to refer to the full lake phase of Lake Mungo, Stage 5e to refer to the last great filling of Lake Eyre in the arid heart of the continent. What had happened, I wondered, while my attention was elsewhere?

The answer was that Emiliani's approach to deep-sea cores had been turned on its head. The post-war period at the University of Chicago was marked by highly strategic research into isotopic chemistry. During this period Willard Libby and

James Arnold developed the use of radiocarbon (^{14}C) — first as a tracer for photosynthetic pathways[5] — and later as the front-line dating technique for archaeology and Quaternary science. This was also the period that Emiliani worked on oxygen isotopes — in this case $^{16}O/^{18}O$ — and their differential incorporation in the calcareous and siliceous skeletons of foraminifer and radiolaria (minute marine protozoa, whose shells form an ooze covering much of the deep ocean floor).[6] Initially, he used the isotopic signatures of the foraminifer to map changes in global sea temperatures. By 1980, researchers at Cambridge had shown that variability in the isotopic composition of the world's oceans was the dominant signal in the deep-sea foraminifer.[7] And that this was substantially a record of ocean volume rather than temperature. The transfer of water vapour from ocean to ice cap selectively removed the light ^{16}O and concentrated the heavier ^{18}O in the oceans. The result was that unassuming benthic (deep sea) foraminifer, rather than titanic glacial moraines, became the building blocks of the global climate record.

The Quaternary appears to have been marked by a repeated cycle of rapid global warming followed by more gradual cooling. The oscillations form a distinctive saw tooth pattern with the warm peaks roughly 100,000 years apart. Cold stages are given even numbers. Warm stages have odd numbers.[8] Today's inter-glacial climate is perched on the peak of one of these saw teeth. Emiliani confidently labelled the modern climate 'Stage 1' — and made no allowance for any *future* climate change. Perhaps scaring the public was unscientific.

The significant feature of the deep-sea record was that it was a continuous record of climatic oscillations and could be correlated with cyclical changes in the Earth's orbit identified by the brilliant Serbian mathematician, Milutin Milankovitch. Cambridge scientists began to describe the deep-sea sediments, as an 'orbital metronome'.[9] Now one could imagine glacials and inter-glacials succeeding each other with each beat. Global climate now began to look precariously balanced. The earlier 1970s view was a primordial rather than processual view, an almost biblical account of Ages. By the 1990s we seemed to be charting the oscillations of an unstable system. But worse was to come.

Millennial mega-flickers

Until a decade ago, even the steepest parts of the marine isotope curve could be imagined as gradual — scarcely perceptible shifts of global climate — taking place over thousands of years. Data from the Greenland and Antarctic ice cores changed that perception. Superimposed on the long-term trends are a series of violent, rapid *flickers* of climate: the *noise* around the long-term trend of the marine isotope curve.[10] From the mid-1990s onwards, papers in the journal *Nature* began to refer to 'sub-millennial' shifts in global climate. Australian National University geomorphologist John Chappell described them as 'millennial mega-flickers'.[11]

The long ice cores provide a fine-grained record of global climate — specifically annual accumulation of snow and ice at the poles over the last 150,000 years.[12] Both the 80-metre Greenland ice core and the 3.6-kilometre long Antarctic 'Vostok' core record a series of sudden temperature jumps in global climate.[13] During the last inter-glacial — the nearest analogue to today's warm climate — switches from warm to cold conditions often took less than a century. Some changes appear to involve shifts of 5–10°C within thirty years. Others show what Quaternarists call 'latching' where the new climate oscillated wildly, then locked in place at one end of the swing for several millennia.[14]

You need to remember that only 5–6°C separates our climate today from that at the height of the last Ice Age. Minus 5°C will see an alpine meadow with button grass transformed into a wilderness of ice and cold, slowly being ground into rock flour under the weight of several kilometres of ice. The implication of the ice core data is that global climate has flickered between glacial and inter-glacial conditions with a rapidity that would not only be perceptible to people, but also catastrophic for any society dependent upon agriculture or engineering works.

All this reinforces the picture of global climate as a fluid, shifting balance between atmosphere, ocean, ice sheets and landmasses — between temperature, precipitation, and solar energy. Small changes build to a threshold or breakpoint — when even a small change may trigger a rapid shift, until a new point of unstable equilibrium is reached. 'All the evidence', says climatologist Jonathan Adams, 'indicates that most long-term climate change occurs in sudden jumps rather than incremental changes'.[15] If that is the case, then catastrophism now seems respectable once again — and gradualism seems dangerously naive. W. S. Broecker, who spent a career attempting to build climate models mimicking the last glacial–inter-glacial cycle, commented that 'climate is an ill-tempered beast, and we are poking it with sticks'.[16]

Climatic assorteds

The full implications of large sub-millennial shifts in climate for Australian biogeography, landscapes and archaeology have barely been considered. There simply has not been time for the data to be absorbed. And there is an existing disciplinary resistance to invoking climate rather than internal dynamics as a causal factor — unless the linkages can be shown. Nearly thirty years ago, biogeographer Donald Walker expressed his unease with routinely invoking past climate change to account for changes in plant and animal communities:

> It is all too easy to become like a child with Christmas chocolates. We open the box of Climatic Assorteds and take our pick of the seductive dainties inside: frosted springs, misty mornings, hot noons, or even glacier mints! Sometimes, with true experimental dash we try something just because we have not tried it before.[17]

El Niño fudge perhaps?[18]

Australian archaeologists have shared this unease. Research into human responses to climate change has become deeply unfashionable. One review of the field commented that 'in this post-modern World,' the archaeology of climate change 'conjures up visions of banal environmental determinism and passive human actors receiving their cues from terrifying landscapes'.[19] A generation of Australian archaeologists broke from older ecological and economic schools of prehistoric research. The trend in current research is to give Aboriginal societies more agency, to focus on the internal dynamics of Aboriginal societies or their social and symbolic dimensions rather than their economic base. This is particularly the case for the Holocene, the last 10,000 years, which in comparison with earlier periods seems an interval of prolonged climatic stability. Many archaeologists doubt that environment has much agency in Aboriginal history during this period. But not everyone agrees. One can hear the frustration in M. J. Rowland's recent comment:

> Space is not just a raw material to be shaped by social process, and landscapes are not merely symbolic constructs. People in the past, as they do today, responded directly to environmental changes, but also indirectly to changes in landscape and resource distribution that were initiated by the changes.[20]

An archaeology of climate change

Two great processes dominate the Quaternary history of Australia: the expansion of the deserts, and the rise of the sea. Between the two lie the temperate fertile crescents of south-western and south-eastern Australia. In response to these changes, the archaeological record preserves a regional history of re-organisation of settlement, repositioning of cultural landscapes, small-scale movement of people in response to the changing configuration of resources, and changes in trade routes and possibly religious networks, and in regional economies.

Rising seas

Rising seas were the most dramatic consequence of post-glacial climate change and entirely transformed coastal landscapes.[21] The transfer of some 50 million cubic kilometres of water from the ice sheets to the oceans raised the sea level by 130 metres, before it stabilised at its present level about 6000 years ago.[22] Whether this rise proceeded at an even rate of 1–1.5 metres per century, or in rushes and sudden rises as ice shelves in Antarctica disintegrated under the impact of global warming, is not clear.[23] But when the sea rose, it flooded about 3,000,000 square kilometres of land, and drowned low-lying land bridges between Tasmania and the mainland, and between New Guinea and Australia — and changed the shape of the continent.[24] In places, where continental shelves were only gently sloping, the loss of coastal territory must have been obvious to people. Calculations for the likely rate of lateral inundation range from 1 metre to over 100 metres per week.[25]

By then, of course, it was not empty land. It was home to various Aboriginal groups, a landscape rich in camps, dreaming sites, histories, places of association, and all the 'landesque capital' — to use Brookfield's phrase — that makes up a productive cultural landscape.[26] All this was drowned by the post-glacial rise in sea level and its cultural consequences have barely been explored. Historian Geoffrey Blainey speculated about the impact:

> Salt water drowned perhaps one-seventh of the land ... Every tribal group on the coast 15,000 years ago must have slowly lost its entire territory ... a succession of retreats must have occurred. The slow exodus of refugees, the sorting out of peoples and the struggle for territories probably led to many wounds and deaths as well as new alliances.[27]

Stirring stuff. But for some archaeologists this emphasised a catastrophist narrative at the expense of the data. Sandra Bowdler argued instead for a 'transliterated' coastal economy, suggesting that the population would have been concentrated along the coastline and would simply have retreated with the coast with minimal disruption.[28] The back country, she argued, would only have been lightly occupied. People would have been pushed back into an empty hinterland. For some parts of the coast this was undoubtedly the case. For others, work since the 1970s has reinforced Blainey's view.

In some places — mainly on sandy coasts — the rising seas destroyed local ecosystems and coastal economies did not re-establish until 3000 years ago, and sometimes not at all.[29] On the Nullarbor coast, the rising seas simply *cut away* any possibility of coastal foraging, carving great sea cliffs into what had been an inland escarpment. The archaeology of Allen's Cave indicates the presence of foragers using the arid open plain as early as 30,000 years ago.[30] Initially the cave was 150 kilometres from the coast. But the sea rose rapidly and by 10,000 years ago was rising up the face of the Nullarbor cliffs only 4 kilometres south of the site. People from coastal districts increasingly used the cave, leaving behind fragments of an abalone shell pendant, and other marine shells; but in this case the translocation of coastal people led to their eventual absorption into the local desert economy.[31]

The flooding of continental shelves also created a score of offshore islands from the former coastal plain. Many of these have strong archaeological signatures reflecting the transition.[32] At Cave Bay Cave, a large sea cave on Hunter Island in Bass Strait, Sandra Bowdler's excavations showed that hunting parties intermittently used the cave around 23,000 years ago when the island 'was a hill on the Bassian plain', the low land that connected Tasmania to the continent.[33] By 7000 years ago, the rising seas had flooded much of the Bass plain, and the cave lay on a peninsula. People using the nearby coast left a dense midden of shells and fish bone in the cave. Then, as Bass Strait took its present form, the island was abandoned.[34]

In some cases, however, people did not leave their country and found themselves marooned by the post-glacial seas. Sooner or later these stranded populations died out. 'Too few people, not enough land, environmental deterioration and no evidence of watercraft during the critical period', explains Bowdler.[35]

In other areas, the 'marine transgression' created new resources and new opportunities. All across northern Australia, shell middens register the creation of extensive mangrove swamps and mangrove forests in the new estuaries around 7000 years ago.[36] In western Arnhem Land, the inundation of inland valleys created a mangrove swamp more than 80,000 hectares in area. Rock paintings provide a remarkable contemporary chronicle of this transformation.[37] The earliest paintings — the delicately detailed 'dynamic figures' — portray life in an arid woodland. Wallabies and other macropods are the most commonly shown animals. Men are painted with elaborate ceremonial headdresses and ornaments, with boomerangs and single-pronged spears. With the rise of the sea, the dominant motifs became estuarine fish — giant perch, barramundi, mullet, catfish — and saltwater crocodile.

This new world was productive but unstable. Even after the seas had stopped rising, morpho-dynamic adjustments of coastal regions continued. From Darwin to Kakadu, the sediment brought down by the coastal rivers gradually choked off the mangrove swamps, leaving large areas of saline mudflats behind a fringe of mangroves along the rivers. As the productive swamps contracted coastward, existing patterns of land tenure, resource use and regional alliances were put under strain. Three thousand years ago, archaeological sites on the floodplains and in the rock shelters show a layer of debris associated with the concerted manufacture of bifacial spear points, a type used for light-weight duelling spears.[38] And the rock art increasingly shows scenes of fighting between large groups of warriors.[39]

Sweepings of the wind[40]

Another consequence of global climate change was expansion and contraction of the Australian desert. Under the cold, dry, windy climate of the last Ice Age, the arid zone expanded dramatically. A vast continental swirl of linear dunes dominates the heart of the continent. The dune fields extended into north-eastern Tasmania, and lie beneath the waters of King Sound on the Kimberley coast.[41]

But unlike the coast, the human history of the desert is more a story of critical resources than transformed landscapes: a history of water with a small 'w'. It is the fluctuating fortunes of wells and soakages, claypan waters and ephemeral lakes that determine access to the country for foragers. These waters are stepping stones through the country. Remove them and access to country and its sparse resources is more limited, the living space of hunters and foragers more confined. In the archaeological record, such changes are reflected in the re-organisation of settlement patterns, changes in residential mobility, changes in the size or configuration of the territory used by a group, and in the foods available.[42]

In the heart of the desert, it took some time to establish a long sequence of prehistoric occupation. In the late 1960s, American archaeologist R. A. Gould reported a 10,000-year record of Aboriginal occupation at Puntutjarpa rock shelter near Warburton.[43] People had settled the desert in more favourable times, argued Gould, and the desert had formed around them as the continent dried out. The desert, he said, was 'a great leveller' of ambition. Adaptation to the limited opportunities of the new desert environment had resulted in a conservative 'Desert Culture' that showed little change or innovation over the following ten millennia. 'Anyone attempting … to adapt in any other way would undoubtedly perish', he said.[44]

This was very much a first approximation to desert prehistory. Later work would show that the desert is a long-standing feature of the Australian continent: it has waxed and waned with global climate change but during the last 50,000 to 60,000 years this has involved changes in degree of aridity rather than transformation on the arid landscape.[45] The full glacial climate in the interior — 20,000 years ago — was drier, windier and colder in winter and hotter in summer than in the same places today.[46] Rainfall is estimated to have been only half of the modern average: 150 millimetres annually, rather than 300 millimetres. Vegetation records for the time indicate scattered trees and sandhill shrubs but little ground cover.[47]

This later work would also show that people had a much longer history of interaction with the desert than first thought, at least twenty millennia earlier than the Warburton dig suggested.[48] This extended human history into the difficult times of full glacial aridity. At Puritjarra rock shelter, west of Alice Springs, people clearly survived and prospered but groups were highly mobile and needed to move over a large territory — perhaps 10,000 square kilometres — to make a living.[49]

I suspect that continuities in both ecology and human settlement depended on rare, extreme rainfall events to recharge wells and soakages, to allow trees to become established and the fauna to re-seed the country. Even during full glacial aridity, there was periodic large-scale flooding. Work on palaeo-floods in central Australia by Mary Bourke shows a large flood event 27,000 years ago. On the Todd River, 'extreme floods have occurred repeatedly since 15,000 BP', with the 'highest magnitude flows between 14,000 and 4000 BP'.[50] The history of the desert is tied up with climatic *variability*, not just climatic *averages*.

On the eastern margin of the arid zone, it was cold rather than aridity that had the most important consequences for people. Lake Mungo — one of a chain of lakes in the Willandra region — filled with glacial runoff from the Snowy Mountains. For more than fifteen millennia this glacial lake was a persistent feature of an otherwise arid landscape, oscillating between freshwater and brackish conditions.[51] People buried their dead on the lakeshore, foraged for shellfish, yabbies, and small marsupials and netted large numbers of fish. Then — around 15,000 years ago — Mungo dried up completely as global warming disrupted the water balance of the lake and reduced discharge down the western rivers. All the large overflow lakes died at this time, beginning with those at the end of the chain and finishing 2000 years later with those closest to Willandra Creek. The people of the Willandra appear to have shifted their focus of activities to the larger river channels and to a handful of small lakes attached to the Murray River, which remained active despite the lower discharge rates.[52] This illustrates some of the resilience of Aboriginal groups in the face of environmental change. Climatic change was not invariably

A senior Martutjarra man walks across newly burnt spinifex country in the Great Sandy desert, 1996

M.A. SMITH

catastrophic. With only small populations, and without significant investment in engineering works and other infrastructure, hunter-gatherer groups had capacity to adjust their focus as the world changed around them.[53]

High fire danger

Archaeologists have tended to give more attention to the impact of people on the environment, than to the effects of global climate change on people. If rising seas and expanding deserts are unequivocal reflections of climate change, there is less consensus on how to untangle the influence of people and climate when we look at changing fire regimes, expansion of grasslands and eucalypt forest and extinction of the large marsupials. Dan Gillespie, attempting to implement joint Aboriginal management of Kakadu National Park in 1981, observed that

> Aboriginal people understand ecosystems like other people wear old overcoats. They know where the thin bits are and know how far they can stretch them before they tear.[54]

The question is, have they ever torn them?

Fire is the usual suspect. Colonial scientists such as Edward Curr in 1888 appreciated its potential as a tool for altering the vegetation:

> There was another instrument in the hands of these savages, which must be credited with results which it would be difficult to overestimate. I refer to the fire stick … he tilled his land and cultivated his pastures with fire; and … almost every part of New Holland was swept by a fierce fire, on an average, once in every five years.[55]

In the 1950s and 1960s, Norman Tindale promoted the idea that the Australian landscape was a product of Aboriginal burning, commenting that 'true primaeval forest may be far less common in Australia than is generally realised'.[56] When the first long vegetation sequences became available as data for study in the early 1980s, there seemed to be some empirical support for this idea. Palynologists such as Gurdip Singh and Peter Kershaw noted that a change from rainforest (or casuarina woodland) to dry eucalypt forest correlated with increased fire frequency — and attributed this change to the long-term effects of people burning the local vegetation.[57] Even in the arid zone — where there are no forests — landscape instability is sometimes attributed to Aboriginal fires. The cumulative effect of Aboriginal burning, concluded one research team, 'has been the mobilisation of vast areas of surface soil resulting in huge sediment loads that have clogged major drainage systems'.[58]

But not everyone is ready to acquit climate change as a suspect. 'The data are being interpreted with the assumption that fire-stick farming is a reality', cautions David Horton, rather than being 'treated as a test of the hypothesis'.[59] 'Climate', says Robin Clarke a researcher who reviewed the fire hypothesis in the 1980s, 'is far more important than fire in determining the distribution of Australian vegetation, but Aboriginal burning might have affected the rate of vegetation change'.[60]

One complication is that any human impact is superimposed on a long-term trend towards replacement of fire-sensitive by fire-promoting vegetation, a trend that reflects increasing aridity of the continent over several million years.[61] Another complication is that wildfire seems to mark major climatic transitions in Australia. 'Any change in vegetation, regardless of whether it was in response to a wetter or drier climate, was likely to be accompanied by an increase in burning', notes Peter Kershaw. 'Vegetation changes are facilitated by fire once climate change has placed existing vegetation under stress.'[62] Bit by bit, people have been reduced to just another source of ignition, with climate setting the parameters of any impact. A key issue seems to be whether or not Aboriginal burning escapes the constraints of local climate, whether burning complements, opposes or amplifies the prevailing climatic trend. Tasmania is a case in point.

Return of the forests

People were using the limestone valleys of south-west Tasmania from 35,000 years ago as the last glaciation waxed and waned around them.[63] This was the extreme edge of the Pleistocene world. One archaeologist summed up the emerging picture: 'extreme climate, extremely low fat diets, extremely rugged country. Why not leave while the Bassian Bridge was intact?'[64] People did not leave, however. For over a thousand generations, small groups of Aboriginal hunters moved through the region, hunting Bennett's wallaby and forest wombats, making bone points and chert scrapers, and bivouacking in small limestone rock shelters for warmth.

Twelve thousand years ago it all came to an end. The major challenge to this way of life was not intense cold but global warming. In the post-glacial period, the glaciers in the higher valleys retreated and dense temperate rain forest colonised much of the region. Archaeologist Jim Allen notes that 'occupation at all sites so far excavated ceased before circa 12,000 BP' and that 'it seems probable that as the encroaching rain forest drove out the game, so did humans abandon the region'.[65] It seems hunters' firesticks could not maintain an open productive grassland, counter to a major climatic trend. The post-glacial return of the forests disrupted Aboriginal settlement across temperate Australia.[66]

A sense of place in time

I have tried to give a partial sketch of the 'geography of the imagination' of archaeologists.[67] We have delved into the space between weather and palaeoclimate without really feeling satisfied with the results. The broad sweep of environmental and climate change has been mapped out by Quaternary science, but it has been hard to make connections with the archaeological record.

Part of the problem has been the temporal and spatial scale of the data. Continental trends give little indication of what the local *on the ground* environmental consequences for human groups might have been. Most archaeological sequences are a conflation of individual occupation events, reflecting very long-term trends, and with a temporal resolution of ±100 years at best and ±500 years more likely. Between the uncertainties of the two data sets, we can barely make out people responding to inter-annual, inter-decadal, or even sub-millennial climate variability.

There is the further difficulty that hunter-gatherer societies tend to be 'soft shelled' and leave little fossil imprint. There are few specialised technologies hostage to specific environmental circumstances. No derelict engineering schemes stranded by climate change. No temple chronicles of drought and famine here. Australian archaeologists have to work with a very coarse-grained historical geography. Our archaeological sequences bear witness to the birth and death of cultural landscapes rather than the finer-grained pulse of cultural ecology. An explicit archaeology of climate change will require more careful attention to research design and modelling at a variety of spatial and temporal scales. And climate will need to become respectable again in archaeological circles.

Despite the problems, deep time narratives have a role in positioning the present. They remind us that human lives are caught up in long rhythms of climate change, and that the present climate is a moment in time. Or, in the words of Thomas Carlyle, a life is but 'a little gleam of time between two eternities'.[68] The record of Quaternary climate change also gives some sense of the dimensions of future climate change — without necessarily providing a close analogue for those changes. Looking at our precarious position at the inter-glacial apex of one of those 'saw teeth', the question seems not whether global climate change will occur — but rather how soon, how fast and how far? The feeling that the long-term dynamics of Quaternary climate were being overlooked in popular debates about global warming led Rhys Jones, archaeologist and iconoclast, to quip:

> Global Warming? Can't get enough.
> Stoke it up higher. Stop the next glaciation!
> *Save civilisation*![69]

M. A. Smith

Michael Smith is an archaeologist and environmental historian and Director, Research and Development, National Museum of Australia, Canberra. He has published extensively on the history and archaeology of Australian deserts, their natural and cultural history, and the history and archaeology of Aboriginal societies from the late Pleistocene to the early twentieth century. He also led the Land and People team that created the *Tangled Destinies* exhibition in the Museum. He has recently co-ordinated a major international collaborative project *23°South*, comparing the archaeology and environmental history of Southern Hemisphere deserts and the prehistory of global colonisation.

16

The social psychology of climate change

Clive Hamilton

According to the world's climate scientists, climate change could see the average temperature of the Earth rise by up to 6°C. During the last Ice Age, when New York was several metres under ice, the mean temperature of the Earth was only 5°C lower than it is now. Sea-level rise of nearly one metre by the end of the century (the upper estimate of the United Nations' Intergovernmental Panel on Climate Change (IPCC)) would see Bangladesh lose 14 per cent of its entire land area and several low-lying islands inundated, inducing a flood of environmental refugees. Tens of millions of people, mostly in poor countries, would be exposed to deadly vector-borne diseases such as malaria, schistosomiasis and dengue.

Quite apart from the implications for the peoples of the Third World, the consequences for Australia of climate change will be horrendous. There is a high probability of many areas settling into a permanent El Niño state. South-west Western Australia will be almost uninhabitable, water flows in the Murray-Darling will fall by 20 to 40 per cent, and severe damage is expected to be visited on the Barrier Reef, Kakadu National Park, the snowfields, wetlands, and so forth.

Yet no one is marching in the streets, and our government refuses to take even small steps to begin to reduce emissions. While in the grip of a drought induced by El Niño, the National Farmers Federation launched a campaign to back the federal government's decision to pull out of the Kyoto Protocol.

What is going on? In a continental nation where the climate deeply affects our psyches, Australians in general, and the government and its policy advisers in particular, appear to be in a state of denial. Opinion polls show widespread concern about the issue, but at another level, Australians don't want to know. George Marshall has analysed the slowness to act in terms of the Freudian idea of denial.[1] Often, when confronted by potential catastrophes, intellectual knowledge is not enough, no matter how compelling it may be. In order to act, people need to be shocked into a state of heightened awareness in which they act on the information in front of them. While some German Jews understood clearly that their lives were imperilled by the Nazis and migrated, many remained in Germany despite the mounting evidence. Marshall quotes an old German adage: 'Things whose existence is not morally possible cannot exist'.[2]

In a way, climate change represents a repudiation of the Enlightenment's faith in the power of science and reason. Humans are capable of a state of simultaneous 'knowing and not-knowing', a state that accurately captures the position on climate change of the US and Australian governments. The Howard government formally accepts the science of climate change and has officially endorsed the reports of the IPCC. For any disinterested reader, the reports of the IPCC make terrifying reading. Yet while ostensibly accepting that all of this is likely to happen, it acts as if nothing will happen. Occasionally, the slip shows and a senior minister will challenge the science, such as when the Prime Minister, confronted with the fears of Pacific island states that they may disappear altogether, said the 'jury is still out' on sea-level rise. In fact, the jury is emphatically in.

The farm lobby is also in denial even though it will be most severely affected. Climate scientists have been saying for some years that global warming is likely to intensify El Niño events, meaning worse droughts across eastern Australia. In fact, climate change represents the most severe threat to the future of farming in Australia. Here are a few facts from a recent CSIRO publication:

- Stream flows in the Murray-Darling Basin are expected to decline by up to 20 per cent by 2030 and 45 per cent by 2070 — nearly a halving of water in a river system already under severe stress — with huge implications for irrigators.
- Annual milk losses of 60 to 90 litres per cow by 2030 are anticipated, even with extra spending on shade sheds and sprinklers.
- Fruit fly and cattle tick are expected to spread southward as average temperatures rise, resulting in loss of export markets.
- Weed infestations will become worse.

All of this seems to have passed by the National Farmers Federation. In fact, at one of its recent meetings a climate scientist was shouted down. The farmers feel so threatened by the news that they won't listen to it. Even the government appears to want to keep the bush in the dark. When the CSIRO released its superb full colour kit called 'Climate Change in Australia', a senior bureaucrat in the Australian Greenhouse Office in Canberra was asked why the kit had not been distributed widely in regional Australia. His response was: 'We don't want to scare the horses'.

Yet the costs to farmers will dwarf the effects of any other issue. A study of the Macquarie River basin in northern New South Wales examined the implications of expected changes in water supplies. The results showed that mean annual runoff will be reduced by up to 30 per cent. Using optimistic assumptions, the study found aggregate losses to the agricultural economy in 2030 of 6 per cent in the low change case and 23 per cent in the high case.

At least one senior rural figure has finally got the message. In 2001, former National Party leader Tim Fischer said: 'There is a very direct challenge for agriculture in New South Wales and Australia arising from climate change, and this will see severe impacts on production over forthcoming decades'. But he is a lonely figure among rural leaders.

Climate change is above all a moral question, because those responsible are rich people who have become rich by burning fossil fuels, and the people who will suffer most are poor people in poor countries. Yet greenhouse deniers — not the mad sceptics in the Lavoisier Group but the Australian politicians, bureaucrats and fossil lobbyists who have prosecuted the Australian position — have attempted to reframe the climate change problem so as to avoid the ethical considerations. The people in question (all of whom I could name) find it too personally challenging to face up to the fact that their own actions jeopardise the lives and livelihoods of multitudes of people. They have attempted to frame the international negotiations as economic or trade issues rather than environmental ones. They have castigated the Europeans for attempting to trick the rest of the world into believing that they are serious about reducing their emissions when in reality they are just trying to obtain kudos for emission reductions that would have happened anyway. Some have even characterised the whole debate as a European plot to win a trade advantage over naive and struggling Australian firms.

At a briefing in Canberra after the Kyoto conference, an Australian Bureau of Agricultural and Resource Economics officer revealed the peculiar worldview that underlay the bureaucratic position, a position shared by the government. Asked to describe the Umbrella Group (which includes Australia, United States, Japan, Canada and Russia), he said it represented the 'free world', as if nine years after the fall of the Berlin Wall the Cold War were still being fought in the Canberra bureaucracy, the European Union having assumed the mantle of the Evil Empire.

It is much harder to make this sort of claim against people in Third World countries, for they are the ones who will suffer disproportionately from the effect of climate change, including decreased crop yields (leading to starvation), sea-level rise, and increased incidence of tropical diseases. The problem of how morally to deal with the innocent victims of climate change caused by us has been solved by attacking poor countries for refusing to save themselves.

Thus after Kyoto, casting around for an excuse to renege on the deal, the Australian government seized on the argument put by the US fossil lobby at the last minute — that the Kyoto Protocol is flawed because developing countries are 'exempted'. This is a manifest denial of the facts and the government's own agreed position, as reiterated by the Prime Minister a week before the Kyoto conference in November 1997. Every international agreement on climate change — the 1992 Framework Convention, the 1995 Berlin Mandate and the 1997 Kyoto Protocol — explicitly recognises that developing countries will be required to cut their emissions, but only after rich countries have led the way. Australia's 20 million people produce more greenhouse pollution than Indonesia's 200 million.

Typically, it was Foreign Minister Alexander Downer who expressed it with his own brand of childish bluntness: '[I]t is no solution at all ... if China and India and Brazil can go ahead and pollute the environment to their heart's content because we're all feeling a bit sorry for them'.[3]

This is a remarkable statement. Blaming the victim is a familiar defence against the fear that it might happen to us. So it is comforting to believe that the unemployed are jobless not because we as a society have failed to provide them with a means of earning a living but because they are lazy bludgers. And we are relieved to hear that a gruesome murder in the next suburb was the result of a 'domestic dispute'. We think: 'As I am not part of the murderer's family, I am safe' and we lose some of our sympathy for the victim. Downer characterises the victims of climate change as selfish — thereby deflecting blame from those who are responsible for the problem — and implies that those who support exclusion of developing countries are soft headed, simultaneously vilifying both the victims and those who support them. He says explicitly that compassion is no basis for action on this issue, a crude attempt to undercut the moral claims of developing countries.

It has been suggested that the reason it was much easier to reach an effective international agreement to phase out ozone-depleting chemicals was that the additional UV rays reaching the Earth through the hole in the ozone layer are especially damaging to the health of people with fair skin. White people, those most affected, tend to be rich and powerful while black people tend to be poor and weak, and the rich and powerful acted because they felt threatened. The story is more complicated, of course. Alternatives to CFCs were available and almost as cheap, and the economies of the world were not dependent on CFCs in the way they are dependent on fossil fuels. But the story does illustrate the way in which threats are interpreted through perceptions of impacts on oneself. This is why in international agreements, developing countries almost invariably come off second best.

Withdrawing from the elements

The need to act on climate change has come up against one of the deepest articles of faith of modern conservative governments — that the principal focus of government should be to maximise the rate of economic growth. Our business and political leaders are so fearful of anything that might affect the growth rate that even a very small price is too much to pay to improve the chances of the long-term survival of capitalist expansion.

Despite the overwhelming evidence of the minimal economic costs of environmental protection and the longer term economic advantages, it remains true that claims that environmental laws will reduce growth have an almost magical power over political decision makers. In the case of global warming, we are confronted by the most frightening threat to the future of the world, set out with chilling understatement in IPCC reports.

As if the predicted impacts I have already referred to — warming, sea-level rise, and millions exposed to deadly diseases — were not daunting enough, in 2002 the US National Academies of Science not only endorsed the IPCC's conclusions but also produced a new report entitled *Abrupt Climate Change: Inevitable Surprises*[4] that argued that global warming may trigger 'large, abrupt and unwelcome regional or global climatic events' such as severe droughts and floods. The smooth curves of the climate models may hide the propensity of the climate system to switch suddenly to new states, with scarcely imaginable results. Even the authors of the National Academies report were concerned that their discussion might curdle the blood of the public and added:

> It is important not to be fatalistic about the threats posed by abrupt climate change. Societies have faced both gradual and abrupt climate changes for millennia and

have learned to adapt through various mechanisms, such as moving indoors, developing irrigation for crops, and migrating away from inhospitable regions.

You can only move indoors once. We are accustomed to securing our homes against some external threats, such as by installing burglar alarms; now there is a literature emerging on how we can 'climate proof' our houses. For modern humans, the weather is increasingly becoming an abstraction, and it is our wealth that enables us to withdraw from the elements. I am reminded of the beautiful Leunig cartoon in which a man and his son marvel at the sunrise on television while the real sunrise occurs at the window. The television image is more real than Nature because we are losing our capacity to join with Nature. Climate proofing our lives is an extension of human attempts to control the natural world and its impacts on us, and this is transforming our psyches in profound ways. As the Earth has filled up with human activity over the last century or two, we harden ourselves against the effects of Nature's unexpected forces, and our relationship to the natural world is transformed. The loss of external wilderness and the isolation of ourselves from the elements — the result of a mere 200 years of industrialism, after tens of thousands of years in which the structure of the human psyche was laid down — have deprived us of the inner wilderness and respect for the natural order.

Without the material counterpart of the primal unconscious, how are we to sustain ourselves psychologically? Under the dual onslaught of economic exploitation and scientific explanation, the world has lost its ability to enchant us. We have nowhere to send Jason on his quest, except perhaps to outer space. In other words, we find it increasingly difficult to find the imagery that can inspire the inner journey through life.

The culture of denial

While conservative governments have found endless excuses for inaction, the people of Australia and the United States have indicated in surveys that they want their governments to take action. But the intensity of concern has been strangely muted. Marshall asks why the general public has been so apathetic on the issue of climate change. Pointing to various psychological processes that reinforce the denial, he suggests that when our grandchildren ask us why we did nothing about climate change even though we understood what would happen, we will either deny knowledge ('I didn't know'), deny our agency ('I didn't do it'), deny our personal power ('I couldn't do anything') or blame others ('The corporations and George Bush did it').

Curiously, some segments of the environment movement in Australia appear to be in denial over climate change too. Although it is unquestionably the most serious environmental threat of all, some environmentalists in Australia have not grasped the gravity of the threat, and we are now seeing a pattern of knee-jerk resistance to a range of renewable energy proposals across this country, from opposition to biomass burning to objections to new wind farms.

Much of the gains of environmental campaigns of the past could be lost if the predictions of the world's climate scientists prove correct. The passion and hard work that has gone into saving the Tasmanian wilderness, protecting the remaining old-growth forests of New South Wales, the tropical forests of Queensland, the jarrah forests of Western Australia and the prodigious efforts that now need to go into restoring the land, all of these may come to nought.

There are important exceptions, notably those groups collected together in the Climate Action Network Australia, along with Greenpeace and the Australian Conservation Foundation. Other environmentalists have been quick to reject proposals for burning biomass, including wood waste, in power plants. Earlier this year, New South Wales environmentalists released a piece of highly dubious research that purported to show that burning forest biomass would result in five times more carbon dioxide than a conventional coal-fired power station. This was an absurd proposition that is contradicted by basic science — while the carbon that goes into coal-fired power plants comes from stable storage underground, the carbon in biomass has been taken out of the atmosphere — yet it has sunk into perceptions in some groups.

More alarming is the local opposition that has developed to wind farms. We have seen it on display with the proposed Portland Wind Energy Project in Victoria. Many groups say they support renewables in principle but they end up opposing it almost everywhere in practice. Even the National Trust seems to have been captured by a general opposition to wind farms. The Trust held a workshop at Deakin University in April 2002. The brochure advertising it gave the impression that there is an industrial juggernaut on the way that has to be stopped, and the speakers list was stacked with opponents of wind farms. The Trust has more recently gone public with its opposition to wind farms. Anyone who has read even cursorily the relevant reports of the CSIRO, not to mention those of the IPCC, will understand that there is no more serious threat to Australia's natural environment than climate change. Apart from anything else, it jeopardises decades of work by the National Trust aimed at conserving Australia's natural heritage.

Australian environmentalism is characterised by parochialism, built on a love for the unique character and importance of the Australian landscape. This has been its strength, but in a world where global environmental problems are to the fore, it is also a weakness. Responses to climate change have been filtered through eyes conditioned to see forests, waterways and landscapes. So proposals to burn wood waste were immediately opposed because they may stimulate logging. Proposals for wind farms have been opposed because they may affect visual amenity and cause bird strike.

Perhaps in Australia, unlike in Europe, the atmosphere is not regarded as part of the environment in the same way as the landscape is. Perhaps this is because in Europe the sky is low and the weather oppressive and encompassing, while in Australia the bigness of the sky invites us to throw out our arms and abandon ourselves to it.

Growth fetishism

We now know what we must do to tackle climate change. We must immediately begin reducing combustion of fossil fuels and keep reducing it until fossil fuels are largely phased out. In 1997, after ten years of hard-fought negotiations, the rich countries of the world agreed to the Kyoto Protocol that would see those countries reduce emissions by around 5 per cent over a decade or so. This would have been a significant first step, yet the agreement instantly came under attack from the fossil fuel lobby in the United States, and the recalcitrant parties managed to insert so many loopholes into the Protocol that, after several international meetings culminating in a conference in Marrakesh in 2001, the Protocol will result in minimal reductions in the greenhouse gas emissions of rich countries over the next decade. Despite the emasculation of the Protocol, soon after his election in 2001 President Bush (whose victory was widely seen to have been engineered by energy giants including Exxon and Enron) declared that doing nothing would be too costly and repudiated the Protocol completely. Australia followed the US lead.

There has been only one reason for the reluctance of the rich countries of the world to reduce their emissions and thereby to stave off environmental catastrophe — the perceived impact of reducing emissions on the rate of economic growth and especially the growth of a handful of powerful industries. This has been enough to jeopardise the future of the world. But on inspection the feared large economic costs of shifting to a low carbon economy prove chimerical. There have been many attempts to estimate the effects of cutting emissions on growth rates using complicated economic models. These models systematically overestimate the negative effects on growth by making a series of assumptions that constrain how businesses can respond to the need to cut emissions.

Yet despite the fact that they systematically overstate the costs of cutting emissions, the economic models consistently produce estimates of reductions in

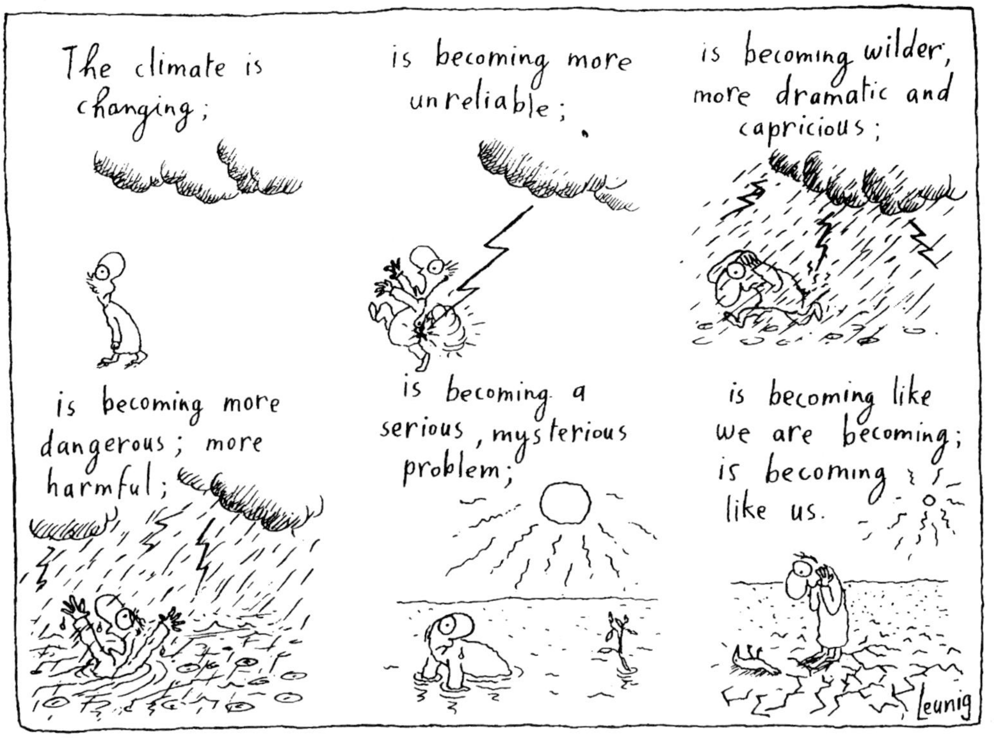

Cartoon by Michael Leunig.

COURTESY OF THE ARTIST AND *THE AGE*

growth rates that are, by any standard, minuscule. They typically conclude that cutting emissions as mandated in the Kyoto Protocol would see the gross national product (GNP) of the United States reduced by half or one per cent by 2010.[5] What does this figure mean? It means that with the required emission reductions, GNP in 2012 is expected to be one per cent lower than it would otherwise have been. But this is a tiny amount. If the economy grows at three per cent per annum over the period then GNP will reach a level around 40 per cent higher by 2012 if nothing is done. According to the models, if policies to reduce emissions as specified in the Kyoto Protocol were implemented, then GNP would be 39 per cent higher by 2012. Put another way, instead of GNP reaching a level 40 per cent higher by, say, 1 June 2012, it will not reach that level until 1 October 2012. The most recent economic modelling studies commissioned by the Australian government show that ratification of the Kyoto Protocol would reduce Australia's GDP by a trifling 0.2 per cent by 2012, a couple of weeks of forgone growth. The effect of the 2002–2004 drought alone on Australian agricultural output will exceed this cost.

Yet in the face of these disappearingly small effects on economic growth, the United States and Australia have refused to play a part in reducing global greenhouse gas emissions. This is growth fetishism taken to an absurd degree. Confronted with a high probability of environmental catastrophe to be visited on the Earth, the richest people on the Earth — people who systematically over-eat and who air-condition the outdoor forecourts of service stations — are unwilling to delay increasing their incomes by 40 per cent by a few weeks.[6] Understood this way, growth fetishism appears to be a form of madness.

It is worth dwelling on the perversity of the application of economic rationality to the global environmental threat posed by climate change. A school of influential economists believes that the enhanced greenhouse effect may not be a bad thing. After all, while some parts of the world will be flooded and others will become intolerably hot, some regions will become wetter and this will improve agricultural productivity. A thorough assessment of the costs and benefits of global warming may well show that the benefits exceed the costs, especially when compared to the alternative scenario, that is, a dramatic reduction in dependence on fossil fuels. It would be entirely consistent with this line of argument to advocate measures to speed up global warming so that we can enjoy the benefits of a hotter planet sooner.

Perhaps the most influential economist advocating this line of argument is William Nordhaus who in a string of papers has analysed the economic impact of the enhanced greenhouse effect on the United States. Nordhaus begins with the most fundamental assumption of economics: 'Whether preventive action should be taken depends on the costs of preventing GHG [greenhouse gas] emissions relative to the damages that the GHGs would cause if they continued unchecked'. Fortunately, Nordhaus reassures us:

> Most of the US economy has little direct interaction with climate, and the impacts of climate change are likely to be very small in these sectors. For example, cardiovascular surgery and microprocessor fabrication are undertaken in carefully controlled environments and are unlikely to be directly affected by climate change.

Nordhaus calculates that only three per cent of national output in the United States is produced in 'climate-sensitive sectors' — mainly agriculture — while another 10 per cent is generated by 'moderately impacted sectors', such as construction, water transportation and energy utilities.

All economic models assess the impact of measures to reduce greenhouse gas emissions in this way, including those used by the Australian government to support its refusal to ratify the Kyoto Protocol. Ordinary mortals despair at the way in which our decisions about the future of the Earth's atmosphere have been captured by the economists' worldview.

The clear implication of this sort of analysis is that economic rationality requires human beings to optimally transform the Earth's climate and any other aspect of the natural environment in order to maximise our welfare. Indeed, Nordhaus finally admits that 'the greenhouse effect might on balance actually be economically advantageous'. This whole line of argument is breathtaking for its faith in humanity's ability to control the natural world.

Of course, with global climate change, we are talking about change to a complete system of unimaginable complexity; a system whose intricacies we have barely begun to understand. Who can possibly predict the consequences of global warming except at the crudest level? And what does it reflect about our respect for the natural environment to imagine that we can regulate the whole to suit our own needs? The belief that twenty-first century humans can manipulate a global ecosystem that has evolved to its present delicate equilibrium over millions of years is perhaps the lowest point of the contempt for Nature that has so characterised the economic worldview of the last two centuries, a vision to which most economists still naively cling.

Clive Hamilton

Clive Hamilton is the Executive Director of the Australia Institute. Before establishing the Institute, he taught economics of development at the Australian National University and conducted policy research in the Australian Public Service including a period as head of research at the Resource Assessment Commission. Clive's research work at the Institute has emphasised climate change policy, measures of wellbeing (especially the Institute's Genuine Progress Indicator), privatisation and taxation issues. Clive is also a Visiting Fellow in the Graduate Program in Public Policy at the ANU and an Adjunct Professor at the University of Technology, Sydney. He has published widely, with seven books including *Running from the Storm: The Development of Climate Change Policy in Australia* (University of New South Wales Press, 2001).

Notes

Human elements

[1] *Argus*, 29 December 1928, p. 15.

[2] Ibid.

[3] *Argus*, 3 August 1935, p. 21.

[4] For this and other descriptions of the difference between weather and climate, see the Bureau of Meteorology's 'Learn About Meteorology' page, online at <http://www.bom.gov.au/lam/>.

[5] For a brief discussion of the definition of climate, see W. J. Gibbs, 'A Perspective of Australian Meteorology, 1939–1978', *Australian Meteorological Magazine*, Vol. 30, 1982, p. 13.

[6] Quoted in T. M. Perry, 'Climate and Settlement in Australia 1700–1930: Some Theoretical Considerations', in John Andrews (ed.), *Frontiers and Men: A Volume in Memory of Griffith Taylor (1880–1963)*, F. W. Cheshire, Melbourne, 1966, p. 141.

[7] For more on concerns about the health effects of tropical climates, see David Walker, 'Climate, Civilisation and Character in Australia, 1880–1940', *Australian Cultural History*, No. 16, 1997/98, pp. 77–95; Warwick Anderson, 'Geography, Race and Nation: Remapping "Tropical" Australia', *Historical Records of Australian Science*, Vol. 11, No. 4, 1997, pp. 457–468; Warwick Anderson, *The Cultivation of Whiteness: Science, Health and Racial Destiny in Australia*, Melbourne University Press, Melbourne, 2002; Neville Nicholls, 'A Healthy Climate?', in Eric K. Webb (ed.), *Windows on Meteorology: Australian Perspective*, CSIRO, Melbourne, 1997, pp. 105–117.

[8] David Laskin, *Braving the Elements: The Stormy History of American Weather*, Doubleday, New York, 1996, pp. 3–4.

[9] *Sydney Morning Herald*, 1 July 1950, p. 2.

[10] Geoffrey Blainey, 'Climate and Australia's History', *Melbourne Historical Journal*, Vol. 10, 1971, pp. 5–9.

[11] Geoffrey Blainey, *A Land Half Won*, Sun Books, Melbourne, 1983, pp. 143–144.

[12] Commonwealth Department of Health and Family Services and Australian Institute of Health and Welfare, *National Health Priority Areas: Report on Cancer Control 1997*, Canberra, p. 24.

[13] Quoted in Richard White, *Inventing Australia*, Allen and Unwin, Sydney, 1981, p. 126.

[14] For more on the role of drought in the formation of national identity, see Brad West and Philip Smith, 'Drought, Discourse and Durkheim: A Research Note', *Australian and New Zealand Journal of Sociology*, Vol. 32, No. 1, March 1996, pp. 93–102; Brad West and Philip Smith, 'Natural Disasters and National Identity: Time, Space and Mythology', *Australian and New Zealand Journal of Sociology*, Vol. 33, No. 2, August 1997, pp. 205–215. For more on familiar media representations of drought, see Asa Wahlquist, 'Media Representations and Public Perceptions of Drought', in Linda Courtenay Botterill and Melanie Fisher (eds), *Beyond Drought: People, Policy and Perspectives*, CSIRO Publishing, Melbourne, 2003, pp. 67–86.

[15] Geoffrey Blainey, *The Tyranny of Distance*, Pan Macmillan, Sydney, 2001.

[16] Geoffrey Blainey, *A Land Half Won*, op. cit., pp. 10–12.

[17] The Bureau of Meteorology has compiled its own catalogue of major climatic events of the twentieth century, 'Drought, Dust and Deluge — A Century of Climate Extremes in Australia', online at <http://www.bom.gov.au/lam/climate/levelthree/c20thc/index20.htm>.

[18] Eric Klinenberg, *Heat Wave: A Social Autopsy of Disaster in Chicago*, University of Chicago Press, Chicago, 2002, pp. 1–11.

[19] Ibid., p. 15.

[20] Ibid., p. 11.

[21] Ibid., pp. 23–24.

[22] Ibid., p. 11.

[23] Emergency Management Australia, 'Heatwaves', online at <http://www.ema.gov.au/ema/emaSchools.nsf/>. See also 'Avoiding Summer's Wave of Heat', *Helix*, No. 69, December–January 2000, online at <http://www.publish.csiro.au/helix/cf/issues/th69a3.cfm>.

[24] Eric Klinenberg, *Heat Wave: A Social Autopsy of Disaster*, op. cit., p. 17.

[25] Tom Griffiths, *Forests of Ash: An Environmental History*, Cambridge University Press, Cambridge, 2001, chapter 10.

[26] Quoted in J. C. Foley, *Droughts in Australia: Review of Records from Earliest Years of Settlement to 1955, Bulletin no. 43*, Bureau of Meteorology, Melbourne, 1957, p. 4.

[27] Janette A. Lindesay, 'Climate and Drought in Australia', in Linda Courtenay Botterill and Melanie Fisher (eds), *Beyond Drought: People, Policy and Perspectives*, op. cit.

[28] Linda Courtenay Botterill, 'Uncertain Climate: The Recent History of Drought Policy in Australia', *Australian Journal of Politics and History*, Vol. 49, No. 1, 2003, pp. 61–74. Recent changes to drought policy are also discussed in Linda Courtenay Botterill and Melanie Fisher (eds), *Beyond Drought: People, Policy and Perspectives*, op. cit.

[29] D. Stehlik, I. Gray and G. Lawrence, *Drought in the 1990s: Australian Farm Families' Experiences*, Rural Industries Research and Development Corporation, Canberra, 1999. See also: Daniela Stehlik, 'Australian Drought as Lived Experience: Social and Community Impacts', in Linda Courtenay Botterill and Melanie Fisher (eds), *Beyond Drought: People, Policy and Perspectives*, op. cit., pp. 87–108; Daniela Stehlik, Andrea Witcomb, Geoffrey Lawrence and Ian Gray, 'Towards a Social Construction of Drought: A Preliminary Analysis', in G. Lawrence, K. Lyons and S. Momtaz (eds), *Social Change in Rural Australia*, Rural Social and Economic Research Centre, Central Queensland University, Rockhampton, 1996, pp. 263–275.

[30] Quoted in D. Stehlik et al., *Drought in the 1990s*, op. cit., p. 56.

[31] Ibid.

[32] Ibid., p. 93.

[33] Brad West, 'Mythologising a Natural Disaster in Post-Industrial Australia: The Incorporation of Cyclone Tracy within Australian National Identity', *Journal of Australian Studies*, No. 66, 2002, pp. 198–204.

[34] Vladimir Jankovic, *Reading the Skies: A Cultural History of English Weather, 1650–1820*, Manchester University Press, Manchester, 2000.

[35] Ibid., p. 11.

[36] For more on the development of meteorological networks in general, see J. Gentilli, 'A History of Meteorological and Climatological Studies in Australia', *University Studies in History*, Vol. 5, No. 1, 1967, pp. 54–79. For more on the development of the Commonwealth Bureau of Meteorology, see the *Metarch* series, available online at <http://www.austehc.unimelb.edu.au/fam/>.

[37] R. W. Home and K. T. Livingston, 'Science and Technology in the Story of Australian Federation: The Case of Meteorology, 1876–1908', *Historical Records of Australian Science*, Vol. 10, No. 2, 1994, pp. 109–127.

[38] *Official Report of the Debates of the Australasian Federal Convention, Adelaide 1897*, Vol. 3, Sydney, 1986, p. 775.

[39] *Commonwealth Parliamentary Debates (Hansard)*, Vol. 32, 1 August 1906, p. 2141.

[40] For more on the links between climate and nationhood, see Tim Sherratt, 'A Climate for a Nation', in *Federation and Meteorology*, Australian Science and Technology Heritage Centre, Melbourne, 2001, online at <http://www.austehc.unimelb.edu.au/fam/0001.html>.

[41] *Argus*, 22 November 1922, p. 9.

[42] Quoted in *Proceedings of the Productivity Conference, Melbourne, August 31 – September 4, 1964*, Bureau of Meteorology, Melbourne, 1964, p. 6.

[43] Ibid., p. 20.

[44] W. M. Prescott, 'The Importance of Weather Forecasts to the Dried Fruits Grower in the Sunraysia Area', in *What is Weather Worth? Papers Presented to Productivity Conference*, Melbourne, August 31 – September 4, 1964, Bureau of Meteorology, Melbourne, 1964, p. 20.

[45] A. R. Johnston, 'The Grazier' in *What is Weather Worth? Papers Presented to Productivity Conference*, Melbourne, August 31 – September 4, 1964, op. cit., p. 17.

[46] For more on continuing tensions between local and centralised weather knowledge, see Andrew Ross, *Strange Weather: Culture, Science and Technology in the Age of Limits*, Verso, London, 1991, pp. 214–215.

[47] D. Stehlik et al., *Drought in the 1990s*, op. cit., p. 19. For more on the power relations embedded in meteorological knowledge, see Andrew Ross, *Strange Weather*, op. cit., pp. 218–219.

[48] Harriet Bulkeley, 'Common Knowledge? Public Understanding of Climate Change in Newcastle, Australia', *Public Understanding of Science*, Vol. 9, No. 3, July 2000, pp. 313–333.

[49] Ibid., p. 313.

[50] Andrew Ross, *Strange Weather*, op. cit., p. 221.

[51] David Laskin, *Braving the Elements: The Stormy History of American Weather*, op. cit., p. 15.

[52] Ibid., p. 5.

[53] Geoffrey Blainey, *A Land Half Won*, op. cit., pp. 350–352; Geoffrey Bolton, *Spoils and Spoilers — Australians Make Their Environment, 1788–1980*, Allen and Unwin, Sydney, 1981, pp. 28–29; Jenny Keating, *The Drought Walked Through: A History of Water Shortage in Victoria*, Department of Water Resources, Victoria, Melbourne, 1992, pp. 41–42.

[54] *Argus*, 29 October 1938, Weekend Magazine, pp. 8–9.

[55] For Jones' own account of his life and work, see Inigo Jones, *My 'Nephelo-Coccygia'*, Brisbane, 1953[?]. For more biographical details, see: John Steele, 'Jones, Inigo Owen (1872–1954)', in Bede Nairn and Geoffrey Serle (eds), *Australian Dictionary of Biography*, Vol. 9, Melbourne University Press, Melbourne, 1983, p. 515; Lennox Walker, *Only an Australian*, published by the author, Golden Beach (Queensland), 1996, pp. 193ff.

[56] Inigo Jones, *My 'Nephelo-Coccygia'*, op. cit., p. 5.

[57] Letter from Jones to Fritz Loewe, 28 January 1949, Fritz Loewe papers, University of Melbourne Archives.

[58] Letter from Alex S. Mortison (Organiser, Inigo Jones Seasonal Weather Forecasting Trust) to Captain J. Francis (MHR), 4 December 1929, National Archives of Australia: A1, 1938/3981.

[59] *Argus*, 7 April 1923, p. 8.

[60] *Argus*, 18 May 1912, p. 19.

[61] For more on Indigenous weather knowledge, see Eric K. Webb, *Windows on Meteorology: Australian Perspective*, CSIRO, Melbourne, 1997, chapters 1–8. The Bureau of Meteorology has recently established an Indigenous weather knowledge site, online at <http://www.bom.gov.au/iwk/>.

[62] Neville Nicholls, 'Cognitive Illusions, Heuristics, and Climate Prediction', *Bulletin of the American Meteorological Society*, Vol. 80, No. 7, July 1999, pp. 1385–1398.

[63] Watkin Tench, '1788', in Tim Flannery (ed.), *Two Classic Tales of Australian Exploration: 1788 by Watkin Tench; Life and Adventures by John Nicol*, Text Publishing, Melbourne, 2002, p. 235.

[64] Geoffrey Blainey, *A Land Half Won*, op. cit., pp. 348–350; Geoffrey Bolton, *Spoils and Spoilers — Australians Make Their Environment*, op. cit., p. 30; Jenny Keating, *The Drought Walked Through: A History of Water Shortage in Australia*, op. cit., pp. 39–40.

[65] Tim Sherratt, 'The Weather Prophets', in *Federation and Meteorology*, Australian Science and Technology Heritage Centre, Melbourne, 2001, online at <http://www.austehc.unimelb.edu.au/fam/0006.html>.

[66] *Argus*, 9 December 1944, Weekend Magazine, p. 2.

[67] *Commonwealth Parliamentary Debates (Hansard)*, House of Representatives, Vol. 6, 31 May 1955, p. 1221.

[68] *Sun-Herald*, 12 May 1957, p. 27.

[69] See, for example: speech by R. L. Chynoweth (Labor MP for Dunkley) on the Murray-Darling Basin Bill, *Commonwealth Parliamentary Debates (Hansard)*, 17 August 1993, p. 51ff; speech by Tony Smith (Liberal MP for Dickson) on Snowy Hydro Corporatisation Bill, *Commonwealth Parliamentary Debates (Hansard)*, 23 October 1997, p. 9752ff.

[70] See, for example: Alan Jones, 'Give Farmers a Hand to Break Awful Cycle', *Daily Telegraph*, 3 October 2002, p. 33; 'Water Figures Large in Nation's Future', *Courier-Mail*, 12 October 2002, p. 24; 'Appeal to Break Cash Drought', *Australian*, 4 October 2002, p. 5.

Climatic outlooks

[1] W. S. Jevons, 'Some Data Concerning the Climate of Australia and New Zealand', in J. W. Waugh (ed.), *Waugh's Australian Almanac for the Year 1859*, James W. Waugh, Sydney, 1859, pp. 47–98; N. Nicholls, 'William Stanley Jevons and the Climate of Australia', *Australian Meteorological Magazine*, Vol. 47, 1998, pp. 285–293.

[2] W. H. L. Ranken, *The Dominion of Australia: An Account of its Foundation*, Chapman and Hall, London, 1874.

[3] N. Nicholls, 'El Niño-Southern Oscillation and Rainfall Variability', *Journal of Climate*, Vol. 1, 1988, pp. 418–421.

[4] C. Todd, in *The Australasian*, 1888, p. 1456.

[5] A. Ångström, 'Teleconnections of Climate Changes in Present Time', *Geografiska Analer*, Vol. 17, 1935, pp. 242–258.

[6] N. Nicholls, 'The Centennial Drought', in Eric K. Webb (ed.), *Windows on Meteorology: Australian Perspective*, CSIRO Publishing, Melbourne, 1997, pp. 118–126.

[7] E. T. Quayle, 'Long-Range Rainfall Forecasting from Tropical (Darwin) Air Pressures', *Proceedings of the Royal Society of Victoria*, Vol. 41, 1929, pp. 160–164.

[8] E. T. Quayle, *On the Possibility of Forecasting the Approximate Winter Rainfall for Northern Victoria*, Bulletin No. 5, Bureau of Meteorology, Melbourne, 1910.

[9] H. M. Treloar, *Foreshadowing Monsoonal Rains in Northern Australia*, Bulletin No. 18, Bureau of Meteorology, Australia, 1934.

[10] A. M. Grant, *The Application of Correlation and Regression to Forecasting*, Meteorological Study No. 7, AGPS, Canberra, 1956.

[11] P. J. Lamb, 'The Climate Revolution: A Perspective', *Climatic Change*, Vol. 54, 2002, pp. 11–28.

[12] F. K. Hare, 'Dynamic and Synoptic Climatology', *Annals of the Association of American Geographers*, Vol. 35, 1955, pp. 152–162.

[13] F. K. Hare, 'The Concept of Climate', *Geography*, Vol. 51, 1966, pp. 99–110.

[14] Ellsworth Huntington, *West of the Pacific*, Charles Scribner's Sons, New York, 1925, pp. 415–416.

[15] K. J. Arrow, 'I Know a Hawk from a Handsaw', in M. Szenberg (ed.), *Eminent Economists: Their Life and Philosophies*, Cambridge University Press, Cambridge, 1992, pp. 42–50; P. L. Bernstein, *Against the Gods: The Remarkable Story of Risk*, Wiley, New York, 1996.

[16] Anonymous, *Report on Meeting on Development of Extended Period Forecasting in the Bureau of Meteorology*, Bureau of Meteorology, Melbourne, 1972.

[17] See R. Allan, J. Lindesay and D. Parker, *El Niño-Southern Oscillation and Climatic Variability*, CSIRO Publishing, Melbourne, 1996, for history of research into the phenomenon, and M. H. Glantz and N. Nicholls, 'A Chronology of Interest in El Niño', in M. H. Glantz, *Currents of Change: Impacts of El Niño and La Niña on Climate and Society*, Cambridge University Press, 2nd edition, 2001, for chronology of interest in El Niño.

[18] A. J. Troup, 'The Southern Oscillation', *Quarterly Journal of the Royal Meteorological Society*, Vol. 91, 1965, pp. 490–506.

[19] Staff Members, Synoptic Research Branch, 'Drought in the Papua New Guinea Highlands During the Period June to September 1972', Working Paper 161, Bureau of Meteorology, Melbourne, 1972; N. Nicholls, 'The Walker Circulation and Papua New Guinea Rainfall', Technical Report 6, Bureau of Meteorology, 1973.

[20] Anonymous, *Report on Meeting on Development of Extended Period Forecasting in the Bureau of Meteorology*, op. cit.

[21] B. Allen and H. Brookfield (eds), 'Frost and Drought in the Highlands of Papua New Guinea', *Mountain Research and Development*, Vol. 9, 1989, pp. 199–334.

[22] A. B. Pittock, 'Climatic Change and the Patterns of Variation in Australian Rainfall', *Search*, Vol. 6, 1975, pp. 498–504; P. B. Wright, 'The Southern Oscillation', in A. B. Pittock et al. (eds), *Climatic Change and Variability: A Southern Perspective*, Cambridge University Press, Cambridge, 1978, pp. 180–184.

[23] B. R. Morton, 'An Applied Mathematician's View', in A. B. Pittock et al. (eds), *Climatic Change and Variability*, op. cit., p. 375.

[24] P. B. Wright, 'The Southern Oscillation', in A. B. Pittock et al (eds), *Climatic Change and Variability*, op. cit., p. 184.

[25] N. Nicholls and F. Woodcock, 'Verification of an Empirical Long-Range Weather Forecasting Technique', *Quarterly Journal of the Royal Meteorological Society*, Vol. 107, 1981, pp. 973–976.

[26] N. Nicholls, 'Air-Sea Interaction and the Possibility of Long-Range Weather Prediction in the Indonesian Archipelago', *Monthly Weather Review*, Vol. 109, 1981, pp. 2435–2443; C. Braak, *Atmospheric Variations of Short and Long Duration in the Malay Archipelago*, Verhandelingen No. 5, Koninklijk Magnetisch en Meteorologisch Observatorium te Batavia, Batavia, Indonesia, 1919.

[27] N. Nicholls, J. L. McBride, and R. J. Ormerod, 'On Predicting the Onset of the Australian Wet Season at Darwin', *Monthly Weather Review*, Vol. 110, 1982, pp. 14–17; N. Nicholls, 'A Possible Method for Predicting Seasonal Tropical Cyclone Activity in the Australian Region', *Monthly Weather Review*, Vol. 107, 1979, pp. 1221–1224.

[28] W. M. Gray, 'Atlantic Seasonal Hurricane Frequency: Part I: El Niño and 30 mb Quasi-Biennial Oscillation Influences', *Monthly Weather Review*, Vol. 112, 1984, pp. 1649–1668; <http://tropicalstormrisk.com>.

[29] N. Nicholls, 'Predictability of the 1982 Australian Drought', *Search*, Vol. 14, 1983, pp. 154–155.

[30] G. L. Hammer, N. Nicholls and C. Mitchell (eds), *Applications of Seasonal Climate Forecasting in Agricultural and Natural Ecosystems*, Kluwer, Dordrecht, 2000.

[31] J. L. McBride and N. Nicholls, 'Seasonal Relationships between Australian Rainfall and the Southern Oscillation', *Monthly Weather Review*, Vol. 111, 1983, pp. 1998–2004.

[32] A. B. Pittock, 'On the Reality, Stability, and Usefulness of Southern Hemisphere Teleconnections', *Australian Meteorological Magazine*, Vol. 32, 1984, pp. 75–82.

[33] N. Nicholls, 'More on Early ENSOs: Evidence from Australian Documentary Sources', *Bulletin of the American Meteorological Society*, Vol. 69, 1988, pp. 4–6.

[34] N. Nicholls, 'How Old is ENSO?', *Climatic Change*, Vol. 14, 1989, pp. 111–115.

[35] T. F. Flannery, *The Future Eaters*, Reed Books Australia, Adelaide, 1994, p. 84.

[36] N. Nicholls, 'How Old is ENSO?', in T. H. Donnelly and R. J. Wasson (eds), *CLIMANZ 3: Proceedings of the Third Symposium on the Late Quaternary Climatic History of Australasia*, Melbourne University, 28–29 November 1987, CSIRO Division of Water Resources, Canberra, 1989, pp. 42–48.

[37] Jared Diamond, *Guns, Germs, and Steel: The Fates of Human Societies*, Norton, New York, 1997, p. 308.

[38] T. F. Flannery, *The Future Eaters*, op. cit., p. 282.

[39] C. J. Limpus and N. Nicholls, 'The Southern Oscillation Regulates the Annual Numbers of Green Turtles (*Chelonia mydas*) Breeding Around Northern Australia' *Australian Journal of Wildlife Research*, 15, 1988, 157–161.

[40] N. Nicholls, 'El Niño-Southern Oscillation Impact Prediction', *Bulletin of the American Meteorological Society*, Vol. 69, 1988, pp. 173–176.

[41] Parliament of Australia, Senate, Hansard report, 20 March 1988.

[42] N. Nicholls, 'Predicting the El Niño-Southern Oscillation', *Search*, Vol. 21, 1990, pp. 165–167.

[43] N. Nicholls and R. J. Allan, 'Locating Early Darwin Pressure Observations: A Marine Meteorological Detective Story', *Marine Studies—A Bulletin of the Marine Studies Centre*, University of Sydney, Vol. 5, No. 19, 1990, p. 2.

[44] R. J. Allan, N. Nicholls, P. D. Jones et al., 'A Further Extension of the Tahiti-Darwin SOI, Early ENSO Events and Darwin Pressure', *Journal of Climate*, Vol. 4, 1991, pp. 743–749.

[45] D. H. White and M. Howden (eds), *Climate and Risk: Agricultural Systems and Information Technology*, Vol. 6, No. 2, Bureau of Resource Sciences, Barton, 1994.

[46] 'El Niño Fails as Reliable Indicator', *Stock and Land*, 18 December 1997.

[47] N. Nicholls, 'Cognitive Illusions, Heuristics, and Climate Prediction', *Bulletin of the American Meteorological Society*, Vol. 80, 1999, pp. 1385–1397.

[48] T. Kestin, 'Variations of Australian Climate and Extremes', PhD thesis, Monash University, August 2000.

[49] J. S. D'Aleo, *The ORYX Resource Guide to El Niño and La Niña*, Oryx Press, Westport, 2002, p. 104.

[50] <http://www.bom.gov.au/bmrc/clfor/cfstaff/nnn/climateinternet/sld001.htm>.

[51] T. J. Trenn and R. K. Merton, 'Descriptive Analysis', in L. Fleck, *Genesis and Development of a Scientific Fact,* University of Chicago Press, 1935 and 1979, pp. 154–165.

Rhythms, patterns, connectivites

[1] J. Cruikshank, 'Glaciers and Climate Change: Scientific Research in Sentient Places', Paper prepared for the American Anthropological Association Annual Meetings, San Francisco, November 2000, p. 5.

[2] Ibid., p. 2.

[3] M. Tumarkin, 'Secret Life of Wounded Spaces: Traumascapes in the Contemporary Australia', PhD Thesis, University of Melbourne, 2002, p. 302.

[4] Greg Holland and John McBride, 'Tropical Cyclones', in Eric K. Webb (ed.), *Windows on Meteorology: Australian Perspective,* CSIRO Publishing, Melbourne, 1997, pp. 200–216.

[5] M. Tumarkin, 'Secret Life of Wounded Spaces', op. cit., p. 251.

[6] W. Caruana, 'Rover Thomas: "Who's that bugger who paints like me?" ' (undated), <www.antiquesandart.com.au/ArticleRover2.html>.

[7] Deborah Bird Rose, *Dingo Makes Us Human: Life and Land in an Australian Aboriginal Culture,* Cambridge University Press, Cambridge, 2000 [1992].

[8] Jane Simpson, 'Perception of Metereology in Some Aboriginal Languages', in Eric K. Webb (ed.), *Windows on Meteorology*, op. cit., pp. 22–23.

[9] R. Hoogenraad and G. J. Robertson, 'Seasonal Calendars from Central Australia', in Eric K. Webb (ed.) *Windows on Meteorology*, op. cit., pp. 34–41.

[10] Ibid., p. 38.

[11] Stephen Davis, 'Documenting an Aboriginal Seasonal Calendar', in Eric K. Webb (ed.), *Windows on Meteorology*, op. cit., pp. 29–33.

[12] Rhys Jones and Betty Meehan, '*Balmarrk Wana*: Big Winds of Arnhem Land', in Eric K. Webb (ed.), *Windows on Meteorology*, op. cit., p. 16.

[13] Rhys Jones, 'Ordering the Landscape', in I. Donaldson and T. Donaldson, *Seeing the First Australians,* George Allen and Unwin, Sydney, 1985, pp. 181–209.

[14] Rhys Jones and Betty Meehan, '*Balmarrk Wana*', op. cit., p. 18.

[15] Dick Kimber, 'Cry of the Plover, Song of the Desert Rain', in Eric K. Webb (ed.), *Windows on Meteorology*, op. cit., p. 12.

[16] For example, see M. Thoms and F. Sheldon, 'Lowland Rivers: An Australian Introduction', *Regulated Rivers: Research and Management*, Vol. 16, 2000, pp. 375–383; K. Walker, J. Puckridge and S. Blanch, 'Irrigation Development on Cooper Creek, Central Australia — Prospects for a Regulated Economy in a Boom-and-Bust Cycle', *Aquatic Conservation: Marine and Freshwater Ecosystems*, Vol. 7, 1997, pp. 63–73; Peter Whetton, 'Floods, Droughts and the Southern Oscillation Connection', in Eric K. Webb (ed.), *Windows on Meteorology*, op. cit., pp. 180–199.

[17] R. Hoogenraad and G. J. Robertson, 'Seasonal Calendars from Central Australia', op. cit.

[18] Barbara Adam, 'Running Out of Time: Global Crisis and Human Engagement', chapter 5 in M. Redclift and T. Benton (eds), *Social Theory and the Global Environment*, Routledge, London, 1994.

Migrants and nomads

[1] I am grateful for the generosity of a number of avian biologists in assisting me with the technical aspects of this chapter, particularly Robert Heinsohn, Henry Nix, Penny Olsen and Julian Reid.

[2] Gilbert White, *The Natural History and Antiquities of Selborne, in the County of Southampton*, T. Bensley for B. White and Son, London, 1789. Preface dated 1 January 1788, and signed Gil. White. This book has been almost continuously in print for the following two centuries, and is the fourth-most republished book in English in the world. See Tom Griffiths, *Hunters and Collectors: The Antiquarian Imagination in Australia*, Cambridge University Press, Melbourne 1996, p. 121.

[3] Gilbert White, *The Natural History and Antiquities of Selborne*, op. cit., p. 36.

[4] R. E. Trebilcock, 'Thursday 18 June', in Siberian Diary, RAOU Archives, SLV MS 9247, p. 19.

[5] R. E. Trebilcock, 'Sunday 21 June', in Siberian Diary, p. 25.

[6] R. E. Trebilcock, 'Sunday 12 July', in Siberian Diary, p. 71.

[7] Robert Hall, 'The Eastern Palæarctic and Australia', *Emu*, Vol. 19, No. 2, October 1919, pp. 82–98; quotes pp. 86, 83. Hall wrote up an earlier account of the expedition in *Ibis*, July 1904, p. 415.

[8] The question of what stimulates breeding in trans-equatorial migrants remained unresolved for many decades. See, for example, Professor Don S. Farner's application to the Fulbright Foundation in 1955 for support for a study of the unknown hormonal stimulants of trans-equatorial migrants and irregular breeding birds, National Archives of Australia, Perth, K1347/1 FARNER 54.

[9] The Australasian Ornithologists' Union was founded in 1901, and became Royal in 1910. In 1996, it changed its name to Birds Australia. See Libby Robin, *The Flight of the Emu*, Melbourne University Press, Melbourne, 2001. Campbell brought out the first edition of his *Nests and Eggs of Australian Birds: Embracing Papers on 'Oology of Australian Birds'* in 1883 (self-published, Melbourne). A larger expanded edition was published in Britain: *Nests and Eggs of Australian Birds including the Geographical Distribution of the Species and Popular Observations Thereon*, Pawson and Brailsford for A. J. Campbell, Sheffield, 1900 and 1901.

[10] Red-necked Avocet *Recurvirostra novaehollandiae*. 'White-headed Stilt' is an older name, based on the scientific name, for the Black-winged Stilt in Australia *Himnatopus himantopus leucocephalus*, which is the white headed race of an international species. Simpson and Day's *Field Guide to the Australian Birds* (sixth edition), p. 347, notes that Red-necked Avocets, Black-winged Stilts and Banded Stilts are the sole representatives of the *Recurvirostridae* (Avocet) family in Australia. See also S. Marchant and P. Higgins (eds), *Handbook of Australian, New Zealand and Antarctic Birds*, Vol. 2, p. 781. All are nomadic, but only the Banded Stilt depends on the arid zone for breeding. Near relations, the plovers (Family *Charadriidae*) are mostly migratory, eight of the thirteen Australian species known being regular migrants from places like Siberia. Jack Jones, 'The Banded Stilt', *Emu*, Vol. 45, No. 1, July 1945, pp. 1–36, reference to Keartland, p. 3.

[11] L. B. Astheimer and W. A. Buttemer, 'Changes in Latitude, Changes in Attitude: A Perspective on Ecophysiological Studies of Australian Birds', *Emu*, Vol. 102, No. 1, 2002, pp. 19–27, especially p. 20.

[12] In the 1930s the 'white headed stilt'. L. Glauert and C. F. H. Jenkins, 'Notes on the Banded Stilt (*Cladorhynchus leucocephalus*) with a Description of its Eggs', *Journal of the Royal Society of Western Australia*, Vol. XVII, 15 January 1931 (read before the Society 14 October 1930), pp. 1–7. In the case of the eggs in Sid Jackson's oological collection, the eggs designated as those of Banded Stilt were neither those of the Banded nor Black-winged Stilt. The same point was made by J. Neil McGilp and A. M. Morgan in 'The Nesting of the Banded

Stilt', *The South Australian Ornithologist*, Vol. XI, No. 2, 1 April 1931, pp. 37–53.

[13] A name confusingly similar to the old vernacular 'White-headed Stilt' used for Black-winged Stilts in the 1930s.

[14] L. Glauert and C. F. H. Jenkins, 'Notes on the Banded Stilt', op. cit., p. 2.

[15] Clive Minton, Grant Pearson and Jim Lane, 'History in the Mating: Banded Stilts Do it Again', *Wingspan*, Vol. 5, No. 2, June 1995, p. 13.

[16] L. Glauert and C. F. H. Jenkins, 'Notes on the Banded Stilt', op. cit., p. 2. The museum's 1930 collection of nineteen successfully blown eggs and a chick was supplemented within a short time by two adult birds in full plumage, an immature bird, and a chick (nesting in down), sent by another Kukerin citizen, Mr W. Broadwith.

[17] A bird is a 'national' of the country in which it is known to breed. So the Siberian breeding waders that spend up to ten months in Australia are just 'visitors', not Australian birds, for those who like to count these things.

[18] I. C. Carnaby, 'The Birds of the Lake Grace District, WA', *Emu*, Vol. 33, No. 2, October 1933, p. 103, rookery information p. 104; I. C. Carnaby, 'A Further Nesting Record of the Banded Stilt', *Emu*, Vol. 46, 1946, pp. 156–158; I. C. Carnaby, 'Nesting Seasons of Western Australian Birds', *The Western Australian Naturalist*, Vol. 4, No. 7, 24 December 1954, pp. 149–156. The 1946 paper records that another 1930 nesting occurred in Lake King, 80 kilometres east of Lake Grace. This was only two years after the Lake King wheat region was first opened up.

[19] I. C. Carnaby, 'The Birds of the Lake Grace District, WA', op. cit., p. 104. There was an unconfirmed report by T. Smith of Kalgoorlie of an abandoned colony of 'Rottnest Snipe' near a lake near Menzies in the spring of 1929. He described a strip of land thirty miles wide 'literally white with dead birds', quoted in L. Glauert and C. F. H. Jenkins, 'Notes on the Banded Stilt', op. cit., p. 2.

[20] He worked closely with Dom Serventy, Jock Marshall and Don Farner on their scientific studies of the physiology of breeding in arid zone birds, as well as himself publishing in *Emu*.

[21] The South Australian Ornithological Association was established in 1899, before the national body, and most of its key early members were enthusiastic eggers. See John Truran, 'The Foundation Period 1899–1933', in R. Collier et al. (eds), *Birds, Birders and Birdwatching 1899–1999*, SAOA, Adelaide, 2000, pp. 18–56.

[22] J. Neil McGilp and A. M. Morgan, 'The Nesting of the Banded Stilt', op. cit., p. 40. 54,000 estimate from Jack Jones, 'The Banded Stilt', [Part II], op. cit., p. 110.

[23] J. Neil McGilp and A. M. Morgan, 'The Nesting of the Banded Stilt', p. 39.

[24] S. Marchant and P. Higgins (eds), *Handbook of Australian, New Zealand and Antarctic Birds*, op. cit., p. 781.

[25] J. Neil McGilp and A. M. Morgan, 'The Nesting of the Banded Stilt', op. cit., p. 41.

[26] 21.3 parts per thousand, according to I. Carnaby, 'A Further Nesting Record of the Banded Stilt', op. cit., p. 158.

[27] J. Neil McGilp and A. M. Morgan, 'The Nesting of the Banded Stilt', op. cit., p. 39. The lake had filled first in February 1930, but there were no informants observing at that time. The Aboriginal people camped in 1918 and feasted on what they described as 'gull's eggs', and although this was not a confirmed sighting, the number of eggs required to feed a group suggested Banded Stilts.

[28] Jack Jones, 'The Banded Stilt', op. cit., in 1945 made the first attempt at remarks on its biogeographical distribution (and drew on the earlier work of Glauert and Jenkins and McGilp and Morgan). See Part I, p. 5, which has a map summarising all observations, breeding and otherwise.

[29] Judith M. Heimann, *The Most Offending Soul Alive: Tom Harrisson and His Remarkable Life*, University of Hawai'i Press, Honolulu, 1999; L. B. Astheimer and W. A. Buttemer, 'Changes in Latitude, Changes in Attitude', op. cit.

[30] D. L. Serventy, 'Biology of Desert Birds', in D. S. Farner and J. R. King (eds), *Avian Biology*, Academic Press, New York, 1971, pp. 287–339, quote p. 292.

[31] Ibid., p. 292.

[32] This change seems to have occurred in the late 1970s. For example, in the first issue of the *Journal of Arid Environments*, J. L. Cloudsley-Thompson, in his editorial, called for articles about the 'problems of the inhabitants of the world's desert regions' (1978, Vol. 1, No. 1, p. 1). In the second editorial, he explicitly mentioned 'intensely ... cold' environments, presumably including those that are not inhabited permanently such as Antarctica (Vol. 2, No. 1, 1979, p. 1).

[33] 'What Are the Arid Lands of Australia?' in website for CSIRO Sustainable Ecosystems: Centre for Arid Zone Research — Alice Springs, <http://www.cazr.csiro.au/aridzone_introduction.htm> (accessed 21 June 2002).

[34] R. J. Newman, 'Pasture Improvement in the Little Desert', *Journal of the Department of Agriculture, Victoria*, Vol. 57, No. 1, January 1959, pp. 1–9. See also Libby Robin, *Defending the Little Desert*, Melbourne University Press, Carlton, 1998.

[35] D. M. Stafford Smith and S. R. Morton, 'A Framework for the Ecology of Arid Australia', *Journal of Arid Environments*, Vol. 18, 1990, pp. 255–278, quote p. 257.

[36] Henry Nix, 'Environmental Determinants of Biogeography and Evolution in Terra Australis', in W.R. Barker and P. J. M. Greenslade (eds), *Evolution of the Flora and Fauna of Arid Australia*, Peacock Publications, Frewville, 1982, pp. 47–66, quote p. 64.

[37] On long-term drying out and the many climatic fluctuations of the Quaternary period in particular, see M.A. Smith, this volume, chapter 15.

[38] J. R. W. Reid, F. J. Badman and S. A. Parker, 'Birds', in M. J. Parker et al. (eds), *Natural History of the North East Deserts*, Royal Society of South Australia Inc., Adelaide, 1990, pp. 169–182. See also R. Schodde, 'Origin, Adaptation and Evolution of Birds in Arid Australia', in W. R. Barker and P. J. M. Greenslade, *Evolution of the Flora and Fauna of Arid Australia*, op. cit., pp. 191–224.

[39] D. M. Stafford Smith and S. R. Morton, 'A Framework for the Ecology of Arid Australia', op. cit., p. 271.

[40] Forty-seven per cent according to J. R. W. Reid, F. J. Badman and S. A. Parker, 'Birds', op. cit. Sources include ephemeral salt lakes, mound springs, artesian bore drains, cattle tanks and troughs.

[41] This puzzled T. Smith of Kalgoorlie in 1930, who observed them dying in thousands even when 'there was water in the lake they came from', quoted in L. Glauert and C. F. H. Jenkins, 'Notes on the Banded Stilt', op. cit.

[42] Quoted in D. L. Serventy, 'Biology of Desert Birds', op. cit., p. 306.

[43] J. A. Keast and A. J. Marshall, 'The Influence of Drought and Rainfall on Reproduction in Australian Desert Birds', *Proceedings of the Zoological Society of London*, Vol. 124, 1954, pp. 493–499.

[44] A. J. Marshall, 'Internal and Environmental Control of Breeding', *Ibis*, Vol. 101, 1959, pp. 456–478, quote

p. 468. But see R. Schodde, 'Origin, Adaptation and Evolution of Birds in Arid Australia', op. cit., who detected weak spring breeding patterns.

[45] Immelmann was only 24 when he arrived in Australia, according to Alec Chisholm's 'Preface' to Klaus Immelmann, *Australian Finches in Bush and Aviary*, Angus and Robertson, Sydney, 1965, p. v.

[46] Klaus Immelmann, 'Drought Adaptations in Australian Desert Birds', in Charles Sibley (ed.), *Proceedings of the Thirteenth International Ornithological Congress*, American Ornithologists Union, Louisiana, pp. 649–657. See also Richard A. Zann, *The Zebra Finch*, Oxford University Press, Oxford, 1996, p. xv.

[47] Marshall died in 1967 and Immelmann in 1987, but much of Immelmann's later work was in ethology rather than ornithology. Both were in their 50s when they died.

[48] Richard A. Zann, *The Zebra Finch*, op. cit.

[49] L. B. Astheimer and W. A. Buttemer, 'Changes in Latitude, Changes in Attitude', op. cit., p. 23. 'Old-endemics' are birds that have long been endemic to Australia, so might be assumed to have evolved their physiological features under earlier climate regimes.

[50] R. Schodde, 'Origin, Adaptation and Evolution of Birds in Arid Australia', op. cit.

[51] J. R. W. Reid and R. P. Jaensch, 'Wetlands in the Desert: How Great is Their Contribution to Waterbird Production and Conservation in Australia and Globally?' Paper presented at the Birds Australia Congress: *Outback Birds: Past; Present; Future,* Alice Springs, 17–18 August 2002.

[52] Joanna Sassoon, 'It ... Lays Eggs Inside a Paper Bag. A Cultural Ecology of Fish Eating Birds in Western Australia', *Environment and History*, Vol. 9, 2003, pp. 31–52.

[53] Reid and Jaensch, 'Wetlands in the Desert' — Abstract, p. 1.

[54] H. J. Frith, *Waterfowl Conservation*, Australian Conservation Foundation Viewpoint Series No. 2, Melbourne, 1968, p. 2.

[55] Fauna and Flora Committee of the Australian Academy of Science, *Proposal to Establish a Biological Survey of Australia*, Australian Academy of Science, Canberra, October 1968 (19 pp. mimeo), pp. 11–12.

[56] C. A. Falzon, 'Nomadism', in David Horton (ed.), *The Encyclopaedia of Aboriginal Australia*, Vol. 2, Aboriginal Studies Press, Canberra, 1994, pp. 794–795.

[57] Ralph Waldo Emerson, 'History', in *Emerson's Essays*, (Introduced by Sherman Paul), J. M. Dent, London, 1955 (first published in this edition 1906), pp. 7–29, quote p. 18.

[58] Ibid., p. 19.

[59] Aldo Leopold, 'Thinking like a Mountain', in *Sand County Almanac and Sketches Here and There*, Oxford University Press, Oxford, 1987 (first published 1949).

[60] This version is slightly different from the Sherman Paul edition. It comes from *The Essays of Ralph Waldo Emerson* (illustrated by John Steuart Curry), Random House, United States, 1944, p. 26.

[61] Ralph Waldo Emerson, 'History', op. cit., p. 28.

[62] Personal observation, August 2002. Arrernte people co-manage this 'biopark' with the National Parks service of the Northern Territory, and many of the visitor guides are local people. See also Deborah Bird Rose, this volume, chapter 3 and R. Hoogenraad and G. J. Robinson, 'Seasonal Calendars from Central Australia', in Eric K. Webb (ed.), *Windows on Meteorology: Australian Perspective*, CSIRO Australia, Melbourne, 1997, pp. 34–41.

[63] Henry Nix, 'Climate Change and its Potential Impact on Australian Fauna and Flora', Birds Australia Congress, 18 August 2002.

[64] Australia signed treaties with Japan (1974) and China (1986) (the JAMBA and CAMBA treaties respectively) to cover the protection of wetlands along the East Asian flyways, the paths taken by international migratory waders. The term 'East Asian' flyway was coined by ornithologist Elliott McClure as part of the Migratory Animal Pathological Survey (MAPS) work by the United States Army in the 1940s. On the treaties, see Libby Robin, *The Flight of the Emu*, op. cit., pp. 306–307.

[65] J. M. Powell, *An Historical Geography of Modern Australia: The Restive Fringe*, Cambridge University Press, Cambridge, 1988, pp. 178–179.

[66] Thomas R. Dunlap, *Nature and the English Diaspora: Environment and History in the United States, Canada, Australia, and New Zealand*, Cambridge University Press, New York, 1999.

Goyder's Line

[1] D. J. Mulvaney, *The Prehistory of Australia*, revised edition, Penguin, Middlesex, 1975, p. 56.

[2] Much of the work on which this paper is based is contained in J. M. Sheldrick, 'Nature's Line: George Goyder over South Australia', PhD thesis, Deakin University, 1999.

[3] [leader], *Advertiser*, 21 January 1882, p. 4.

[4] 'Earth Hunger', *South Australian Register*, 1 March 1872, p. 5. Goyder is quoted from Parliamentary Paper 28, South Australia, Parliament 1888, *Report of the Commission on the Land Laws of South Australia*, minutes, p. 77, question 2201.

[5] South Australia, Parliament 1890, *Report on the Disposal of Crown Lands*, Parliamentary Paper 60, pp. 16–17.

[6] K. R. Bowes, 'Land Settlement in South Australia 1857–1890', PhD thesis, Australian National University, 1963, p. 197.

[7] Parliamentary Paper 60, South Australia, Parliament 1890, *Report on the Disposal of Crown Lands*, p. 16.

[8] 'Song of Praise', *Pasquin*, 24 October 1868, and Debates in the Legislature, South Australia, Parliament, 22 January 1882, c. 171. R. W. E. Henning is quoting a letter from Friedrich Kuttler of Winnowie.

[9] H. A. Lindsay, 'Surveyor Goyder — of the Rainfall Line', *Riverlander*, September 1963, p. 33.

[10] Parliamentary Paper 72, South Australia, Parliament 1857–1858, *Northern Explorations*, p. 4.

[11] 'The Exploration of the Interior', *South Australian Register*, 30 July 1857, p. 2.

[12] Parliamentary Paper 174, South Australia, Parliament 1857–1858, *Country Adjacent to Lake Torrens*, p. 1.

[13] J. E. Tenison-Woods, *A History of the Discovery and Exploration of Australia*, H. T. Dwight, Melbourne, 1864, pp. 244–250 and 'Rivers of Interior Australia', *Register*, 28 January 1864, p. 2.

[14] Parliamentary Paper 177, South Australia, Parliament 1860–1861, *Mr. Goyder's Report on Northern Triangulation*, p. 3.

[15] Parliamentary Paper 33, South Australia, Parliament 1891, *Report of the Pastoral Commission*, minutes, p. 79, question 2043.

[16] Lionel C. E. Gee, 'The South-East Fifty Years Ago, III, Vanishing Aborigines', *South Australian Register*, 24 March 1928, p. 5.

[17] Parliamentary Paper 20, South Australia, Parliament 1860, *Geological Notes* by A. R. C. Selwyn, p. 1.

[18] Parliamentary Paper 82, South Australia, Parliament 1865–1866, *Report of Surveyor-General on Northern Runs*, p. 1.

[19] Parliamentary Paper 22, South Australia, Parliament

1902, *Report of the Royal Commission on the Pinnaroo Railway*, minutes, p. 65, question 2964.

[20] South Australia, Parliament, Debates in the Legislature, 22 January 1882, c. 173.

[21] Parliamentary Paper 72, South Australia, Parliament 1865, *Report of the Select Committee of the House of Assembly ... [on the] System of Selling the Crown Lands*, minutes, p. 72, question 2200, and also p. 70, questions 2150–2169.

[22] Parliamentary Paper 72, op. cit., p. 75, questions 2277–2278.

[23] From Parliamentary Paper 154 of 1865–1866. The lighter segments of line at the top are not related. The shaded area in the south represents sold agricultural land. By 1880 wheat farming extended north above the peak of the line beyond this section of map.

[24] South Australia, Parliament, Debates in the Legislature, 29 September 1858, c. 260.

[25] 'The Drought in the North', *South Australian Register*, 23 June 1865, p. 2.

[26] Parliamentary Paper 62, South Australia, Parliament 1865–1866, *Instructions to Surveyor-General.*

[27] Parliamentary Paper 78, South Australia, Parliament 1865–1866, *Surveyor-General's Report on Demarcation of Northern Rainfall.*

[28] Parliamentary Paper 82, South Australia, Parliament 1865–1866, *Report of Surveyor-General on Northern Runs*, p. 2.

[29] Parliamentary Paper 133, South Australia, Parliament 1856–1866, *Memo of Surveyor-General on Rainfall*, & c.

[30] *Plains of Contrast: A History of Willowie-Amyton-Booleroo-Whim District*, Centenary Book Committee, South Australia, 1975, p. 5.

[31] *Pastoral Pioneers of South Australia*, Vol. 1, Publishers Ltd, Adelaide, p. 129.

[32] *Mr. Goyder's Report on Victoria Land Regulations*, *Advertiser*, Adelaide, 1870, pp. 7–8. [Also as Parliamentary Paper 23, South Australia, Parliament 1870–1871, *Report on Victoria Land Regulations.*]

[33] Anthony Trollope, ed. P. D. Edwards and K. B. Joyce, *Australia*, University of Queensland Press, St Lucia, 1967, pp. 656–657. (First published 1873.)

[34] Parliamentary Paper 28, South Australia, Parliament 1888, *Report of the Commission on the Land Laws of South Australia*, minutes, p. 77, question 2200.

[35] [leader], *Advertiser*, 21 January 1882, p. 4.

[36] Parliamentary Paper 109, South Australia, Parliament 1882, *Annual Progress Report upon State Forest Administration, 1881–1882*, p. 21, paragraph 224.

[37] Office of the Surveyor-General, *The Past and Present Land Systems of South Australia*, Government Printer, Adelaide, 1881, p. 10. Goyder is evidently the anonymous author.

[38] T. Griffith Taylor, *The Australian Environment (Especially as Controlled by Rainfall)*, Advisory Council of Science and Industry, Melbourne, 1918, p. 98.

[39] Donald W. Meinig, 'Goyder's Line of Rainfall: The Role of a Geographic Concept in South Australian Land Policy and Agricultural Settlement', *Agricultural History*, Vol. 35, October 1961, pp. 207–214, p. 207 and p. 210.

[40] 'Death of Mr. G. W. Goyder, C. M. G.', *Adelaide Observer*, 5 November 1898, p. 32.

Rainmaking in CSIRO

[1] National Archives of Australia (NAA), A8689/1/1945/1.

[2] CSIR, 21st *Annual Report (for the Year ended 30th June, 1947)*, p. 7.

[3] C. B. Schedvin, *Shaping Science and Industry: A History of Australia's Council for Scientific and Industrial Research, 1926–49*, Allen and Unwin, Sydney, 1987, pp. 311–315.

[4] When Fred White joined the CSIR Executive, he was succeeded as chief of Radiophysics by the wartime deputy chief, J. N. Briton. This, however, was always seen as a short-term arrangement since Briton had made it clear that, when peace came, he intended returning to the position in the radio industry from which he had been seconded for the duration of the war to work at Radiophysics. White had no doubt that in the longer term Bowen should become chief, and Bowen had already been offered the position at the time he prepared the first of his reports. White and Bowen had known each other since the early 1930s when they were fellow members of the radio research group in the Physics Department at King's College London — White as an assistant lecturer and Bowen as a PhD student — and White had later been responsible for Bowen's being seconded to Australia in 1943 from the British radar project.

[5] E. G. Bowen, 'Future Programme of the Division of Radiophysics', agenda item for 35th Session of CSIR Council, 2 July 1945, CSIRO Radiophysics archives.

[6] Other radio research groups were also actively concerned with the links between radio propagation and meteorology. See, for example, *Meteorological Factors in Radio-wave Propagation: Report of a Conference Held on 8 April 1946 at The Royal Institution, London*, Taylor and Francis for The Physical Society, London, 1946.

[7] E. G. Bowen, 'Programme of the Division of Radiophysics', 8 August 1946, CSIRO Radiophysics archives, D.1/2. Bowen also prepared an interim report, in May 1946, for Fred White; see E. G. Bowen to F. W. G. White, 23 May 1946, CSIRO Radiophysics archives, D.1/2.

[8] E. G. Bowen to F. W. G. White, 23 May 1946, op. cit.

[9] F. W. G. White, 'Radiophysics Laboratory', 14 December 1945, NAA, A8689/1/1945/1.

[10] J. L. Pawsey, R. Payne-Scott and L. L. McCready, 'Radio-frequency Energy from the Sun', *Nature*, Vol. 157, 1946, pp. 158–159.

[11] Woodruff T. Sullivan III, 'Early Years of Australian Radio Astronomy', in R. W. Home (ed.), *Australian Science in the Making*, Cambridge University Press, Cambridge, 1988, pp. 308–344.

[12] The attempt was, however, occasionally made. For example, in CSIRO's 1949 *Annual Report*, p. 105, we read that 'attempts to explain how certain types of radio waves arise in the Sun are already leading to new techniques for the generation and amplification of radio waves'.

[13] E. Kraus and P. Squires, 'Comments on the Possibility of Rain Stimulation', 5 December 1946; NAA (New South Wales), C3830/15, Box 74.4.

[14] CSIR, 21st *Annual Report for the Year ended 30th June 1947*, p. 98.

[15] Brad Collis, *Fields of Discovery: Australia's CSIRO*, Allen and Unwin, Sydney, 2002, pp. 368–371. See also Brian Ryan, 'Cloud Physics Research 1949–1984 and Beyond', in Eric K. Webb (ed.), *Windows on Meteorology: Australian Perspective*, CSIRO Publishing, Melbourne, 1997, pp. 142–161.

[16] The importance to the Radiophysics Division of having an 'applied' strand to its investigations, to counteract the weight of effort being put into 'pure' radio astronomy research, can be seen again in the 1970s and 1980s, following the transfer of the cloud physics work out of the Division, in the enormous effort put into developing the 'Interscan' microwave landing system for the international aviation industry.

[17] CSIRO, *Eighth Annual Report*, 1955–1956, p. 129. A

major American publication declared, some years later, that the CSIRO group 'could probably be called the world's most outstanding scientific group dealing with cloud physics and weather modification'; Horace R. Byers, in W. N. Hess (ed.), *Weather and Climate Modification*, John Wiley and Sons, New York, 1974, p. 23.

[18] CSIRO, *First Annual Report for the Year ended 30th June 1949*, p. 104.

[19] CSIRO, *Second Annual Report for the Year ended 30th June, 1950*, p. 120.

[20] B. F. Ryan and W. D. King, 'A Critical Review of the Australian Experience in Cloud Seeding', *Bulletin of the American Meteorological Society*, Vol. 78, 1997, pp. 239–254, quote p. 241. See also E. J. Smith, 'Cloud Seeding in Australia', in W. N. Hess, *Weather and Climate Modification*, op. cit., pp. 432–453.

[21] E. G. Bowen, 'The Influence of Meteoritic Dust on Rainfall', *Australian Journal of Physics*, Vol. 6, 1953, pp. 490–497.

[22] *Australian Journal of Physics*, Vol. 7, 1954, pp. 354–358 and pp. 358–364 respectively.

[23] A. C. B. Lovell to E. G. Bowen, 22 December 1953; NAA (New South Wales), A1/11/23.

[24] T. W. Davidson and D. F. Martyn, 'A Supposed Dependence of Meteor Rates on Lunar Phase', *Journal of Geophysical Research*, Vol. 69, 1964, pp. 3981–3987; D. F. Martyn, 'Some Statistical Pitfalls in the Study of Geophysical Time Series', *Australian Journal of Science*, Vol. 27, 1964, pp. 249–252.

[25] B. F. Ryan and W. D. King, 'A Critical Review of the Australian Experience in Cloud Seeding', op. cit.

[26] 'Statement by Director of Meteorology on Cloud Seeding', 27 April 1965, NAA (New South Wales), C3830/13/Z1/38.

[27] 'Cloud Seeding Research Ends', CSIRO news release, 28 October 1981, NAA (New South Wales), C4662/1, Box 49.3.

[28] C. B. Schedvin, 'Atmosphere, Clouds and Rainmaking', unpublished draft, p. 23.

[29] E. G. Bowen to J. R. Price, 23 June 1970; NAA (New South Wales), C3830/13/Z1/57.

[30] 'CSIRO Study Refutes Claimed Regional Effects of Cloud Seeding', CSIRO news release, 19 August 1983, NAA (New South Wales), C4662/1, Box 49.3.

[31] B. F. Ryan and W. D. King, 'A Critical Review of the Australian Experience in Cloud Seeding', op. cit.

Managing climate

[1] Donald W. Meinig, *On the Margins of the Good Earth*, Rigby, Adelaide, 1970.

[2] Interestingly, among many other things, the Royal Commission recommended a 'total watershed approach', something that is still advocated but has not yet been achieved. Its report also provided strong arguments based on North American advice and experience to the effect that water licences should be tied to land ownership to avoid the dangers of water monopolies.

[3] J. M. Powell, *Watering the Garden State*, Allen and Unwin, Sydney, 1989, pp. 137–143.

[4] George Reid, *My Reminiscences*, Cassell and Co., London, 1917, p. 136.

[5] *Irrigation in Western America* — First progress report of the Royal Commission into Water Supply, Victorian Parliamentary Papers, 1885, p. 56.

[6] Parliament of Victoria, *Interstate Royal Commission on the River Murray*, Melbourne, 1902, p. 15.

[7] Murray-Darling Basin Agreement, Part X, Division 1 — State Entitlements to Water, 1992, pp. 44–48.

[8] Wolf Aaron, 'Transborder Waters: Sharing Benefits, Lessons Learnt', Paper delivered at the International Conference on Freshwater, Bonn, 2000.

[9] *The Murray-Darling Basin Salinity Audit*, Murray-Darling Basin Commission, Canberra, 1999.

[10] G. R. Walker, M. Gilfedder and J. Williams, *Effectiveness of Current Farming Systems in the Control of Dryland Salinity*, CSIRO Land and Water, Canberra, 1999.

[11] Ibid., pp. 7–13.

[12] *Murray Valley Salinity Investigation*, River Murray Commission, a report prepared by the consultants Gutteridge, Haskins and Davey, especially Chapter 9 'The Supply Channel Scheme', 1970, pp. 208–225.

[13] *The Murray-Darling Basin Salinity Audit, Snapshot of the Murray-Darling Basin River Condition*, op. cit., Murray-Darling Basin Commission, Canberra, 2001; *The Living Murray*, Murray-Darling Basin Commission, Canberra, 2002.

[14] *Greenhouse Sinks and the Kyoto Protocol*, Australian Greenhouse Office, Canberra, 2000, p. 49.

[15] Roger Jones et al., *Future Impacts of Climate Variability, Climate Change and Land Use Change on Water Resources in the Murray-Darling Basin*, CSIRO Atmospheric Research, September 2001.

[16] Ibid.

[17] *The Living Murray*, op. cit.

[18] *Review of the Cap*, Murray-Darling Basin Commission, Canberra, 2000; *The Living Murray*, op. cit.

[19] *Water Reform Waters Down Farmers Rights*, National Farmers Federation, Media Release 31 October 2001.

[20] See the *Independent Report of the Expert Reference Panel on Environmental Flows and Water Quality Requirements for the River Murray System*, Murray-Darling Basin Commission, Canberra, 2002. The report argues that restoration of 1,950 gigalitres would give a moderate probability of recreating a healthy River Murray system and restoration of 4,000 gigalitres, a high probability. Diversions from the Murray system currently average approximately 8,500 gigalitres (this excludes those rivers that flow into the Darling River). These two options are known as D and E respectively, and are not being discussed in the community consultation process.

[21] *The Water Audit*, Murray-Darling Basin Commission, Canberra, 1995.

The curse of the tropics

[1] Francis Hopkins, *Reaping the Whirlwind: An Australian Patriotic Drama for Australian People*, Websdale and Shoosmith, Sydney, 1909.

[2] *Northern Territory Acceptance Act 1910, reprinted as at 31 December 1983*, Canberra, 1984.

[3] Henry Thomas Buckle, *History of Civilization in England*, 3 volumes, Grant Richards, reprinted, London, 1903–1904, pp. 30–113.

[4] Cited in 'White Races in the Tropics', *Japan Times*, 9 August 1908.

[5] Gitta Sereny, *Albert Speer: His Battle with Truth*, Picador, London, 1996, p. 252.

[6] Henry Thomas Buckle, *History of Civilization in England*, op. cit., Vol. 1, pp. 93–97.

[7] Benjamin Kidd, *The Control of the Tropics*, Macmillan, New York, 1898, pp. 51–52.

[8] Vilhjalmur Stefansson, *The Northward Course of Empire*, George G. Harrap and Co. Ltd, London, 1922. See graph opposite title page.

[9] Ellsworth Huntington, *Civilization and Climate*, Yale

University Press, New Haven, 1915. Similar views persisted in accounts of racial aptitudes. For the view that the Japanese were a people of tropical origin saved from indolence by the severity of their winter climate, see Ian Morrison, *This War against Japan: Thoughts on the Present Conflict in the Far East*, Faber and Faber, London, 1943, p. 11.

[10] John Foster Fraser, *Australia: The Making of a Nation*, Cassell and Company Ltd., London, 1910, pp. 52–63.

[11] Griffith Taylor, 'Our Tropical Spaces', *Forum*, 7 November 1923; 'Tropical Problems: Climate, Health and Settlement', *Argus*, 14 October 1922; 'The Australian Tropics', *News Bulletin*, Institute of Pacific Relations, January 1927.

[12] Griffith Taylor, 'Australia's Future Population', *The Morpeth Review*, No. 4, June 1928.

[13] Patrick Joseph Lynch [Senate], *Northern Territory Acceptance Act*, 21 September 1910, p. 3470.

[14] John Foster Fraser, *Australia: The Making of a Nation*, op. cit., p. 214; Richard Arthur, 'The Question of Questions for Australia', *Australasian Review of Reviews*, n.d. Richard Arthur Papers, MS 473, Mitchell Library, Sydney.

[15] Ambrose Pratt, *The Big Five*, Ward, London, 1910, pp. 180–194.

[16] Gizen-No-Teki [E. W. Foxall], *Colorphobia: An Exposure of the 'White Australia' Fallacy*, R. T. Kelly, Sydney, 1903, p. 3.

[17] Charles Edward Woodruff, *Expansion of Races*, Rebman, London, 1909, pp. 277–278.

[18] Hardy Wilson, *The Dawn of a New Civilization*, Palmer, London, 1929, p. 95.

[19] A. Grenfell Price, *White Settlers in the Tropics*, American Geographical Society, New York, 1939.

[20] See George E. Morrison, *An Australian in China*, first published London, 1895, Oxford University Press reprint, Melbourne, 1985, p. 223.

[21] W. M. Hughes, *The Splendid Adventure: A Review of Empire Relations within and without the Commonwealth of Britannic Nations*, Ernest Benn Ltd, London, 1929, p. 369.

[22] Ibid., p. 15.

[23] Randolph Bedford, *Explorations in Civilization*, S. Day, Sydney, 1916, p. 9.

[24] Ibid., p. 173.

[25] Roger Benjamin (ed.), *Orientalism: Delacroix to Klee*, The Art Gallery of New South Wales, Sydney, 1997.

[26] This widely held nineteenth-century view is discussed in Donald W. Meinig, *On the Margins of the Good Earth: The South Australian Wheat Frontier*, 1869–1884, Rigby Ltd, Adelaide, 1962, pp. 59–60.

[27] See Bradley Fitzmaurice, 'Red Parasols over an Antipodean Italy: *The Lone Hand* and the People Who Made it', MA thesis, University of New South Wales, 1986.

[28] Frank Fox, *Australia*, A and C Black Ltd, London, 1927 (first published 1910), dedication.

[29] Michael Roe, *Nine Australian Progressives: Vitalism in Bourgeois Social Thought 1890–1960*, University of Queensland Press, Brisbane, 1984. See also Alison Bashford, 'At the Border: Contagion, Immigration, Nation', *Australian Historical Studies*, Vol. 33, October 2002, pp. 344–358.

[30] Warwick Anderson, *The Cultivation of Whiteness: Science, Health and Racial Destiny in Australia*, Melbourne University Press, Melbourne, 2002, pp. 127–152.

[31] See Ross Laurie, ' "Not a Matter of Taste but a Healthy Racial Instinct". Race Relations in Australia in the 1920s: Racial Ideology and the Popular Press', MA thesis, Griffith University, 1989, p. 51.

[32] Penny Russell, 'Recycling Femininity: Old Ladies and New Women', *Australian Cultural History* (special issue on 'Bodies'), No. 13, 1994, pp. 31–51.

[33] Theodore Zeldin, *France 1848–1945: Intellect and Pride*, Oxford University Press, Oxford, pp. 105–106.

[34] Leon Cantrell (ed.), *A .G. Stephens: Selected Writings*, Angus and Robertson, Sydney, 1978, pp. 58–64.

[35] Charles Daley, 'Social Life in Australia: Its Advantages and Outlook', *Australia To-Day*, 10 December 1908.

[36] Philip E. Muskett, *The Attainment of Health and the Treatment of the Different Diseases by Means of Diet*, William Brooks and Co. Ltd, Sydney, n.d., p. 4.

[37] Paul Fussell, *Abroad: British Literary Traveling Between the Wars*, Oxford University Press, New York, 1980.

[38] Arthur Adams, *The Australians*, Nash, London, 1920, p. 30.

[39] Ellsworth Huntington, *West of the Pacific*, Charles Scribner's Sons, New York, 1925, p. 340.

[40] 'The Man of the Week: Counsel for the Yellow Streak', *Smith's Weekly*, 14 July 1923.

[41] Australasian Medical Congress, *'Tropical Australia': Transactions of the Eleventh Session, Brisbane, 21–28 August 1920*, Brisbane, 1921.

[42] Arthur Adams, *The Australians*, op. cit., p. 30.

[43] See also E. J. Brady, *The Land of the Sun*, Edward Arnold and Co., London, 1924, p. 12.

[44] Raphael Cilento, *The White Man in the Tropics: With Especial Reference to Australia and its Dependencies*, Department of Health Service, Melbourne, 1925.

[45] See chapter 12: 'Space Becomes Place', David Walker, *Anxious Nation: Australia and the Rise of Asia, 1850–1939*, University of Queensland Press, Brisbane, 1999.

[46] Warwick Anderson, *The Cultivation of Whiteness*, op. cit., pp. 57–61.

[47] Raphael Cilento and Clem Lack, *Triumph in the Tropics: An Historical Sketch of Queensland*, Smith and Patterson, Brisbane, 1959.

'All just melted with heat'

[1] Isabella Wyly to her sister, 6 March 1857, quoted in David Fitzpatrick, *Oceans of Consolation: Personal Accounts of Irish Migration to Australia*, Melbourne University Press, 1995, p. 117.

[2] Warwick Anderson, *The Cultivation of Whiteness: Science, Health and Racial Destiny in Australia*, Melbourne University Press, Melbourne, 2002.

[3] [J. William] Mackenna, 'On the Mortality of Children', *Australian Medical Journal* [*AMJ*], 1858, pp. 119–141, quote p. 123.

[4] *AMJ*, 1858, pp. 222–231.

[5] This larger project: 'Birth Weight, Social Class and Life Courses: The Health and Social Destinies of Babies Born at the Women's Hospital, Melbourne, 1857–1936' is being conducted by the author in collaboration with Dr Ruth Morley, VicHealth Fellow, Department of Paediatrics, Royal Children's Hospital, Melbourne, and Dr Shurlee Swain, Australian Catholic University.

[6] *Victorian Police Gazette*, 12 January 1875.

[7] Colony of Victoria, Records of the Royal Park Receiving House, 1880–1888, No. 12656 (Microfiche version).

[8] Death certificate, 1875, 26 February, 4215, at Richmond. Lying-in Hospital Midwifery Book, Royal Women's Hospital Archives, (LIH)10 October 1874, Mary Jane Patching, 28, married, born England, para 4, 12 hours labour, 8 pounds, 20 inches, forceps and chloroform.

[9] Temperatures courtesy of the Bureau of Meteorology.
[10] Death certificate, 1875, 19 January, 3044, at Madeline Street, Carlton; LIH 23 July 1874, 22, born Victoria, para 2, single, 3 hours labour, 6 pounds, 19 inches.
[11] Death certificate, 1875, 22 January, 3297, at Curzon Street, Hotham; LIH 9 July 1874, 27 born Ireland, para 1, single, 12 hours labour, 6 and a half pounds, 18 inches.
[12] Death certificate, 1875, 22 January, 1190, at Napier Street, Fitzroy, diarrhoea and convulsions for 6 days; LIH 7 October 1874, 22, married, born Ireland, para 1, 6 hours, 6 pounds, 19 inches.
[13] Death certificate, 1875, 24 January, 3072, at Lygon Street, dentition diarrhoea for 2 months; LIH 8 February 1874, 24, married, born England, para 2, 3 hours, 6 and a half pounds, 19 inches; Births, Deaths, Marriages Victoria.
[14] Colony of Victoria, Children's Register, 1880–1888, 14282, microfiche.
[15] *AMJ*, 15 December 1882, pp. 556–558.
[16] *AMJ*, November 1876, p. 367.
[17] Mike Davis, *Late Victorian Holocausts: El Niño Famines and the Making of the Third World*, Verso, London and New York, 2001, pp. 254–256.
[18] See Janet McCalman and Ruth Morley, 'Mothers' Health and Babies' Weights: The Biology of Poverty at the Melbourne Lying-In Hospital, 1857–1883', *Journal of the Social History of Medicine*, Vol. 16, No. 1, 2003, pp. 39–56.
[19] Ibid.
[20] Death certificate, 1875, 17354. No trace of Eveline Greville, the mother's maiden name given on the death certificate, nor of Eveline Delmer/Delmar can be found in the *Tasmanian Pioneers Index*.
[21] Death certificates, 1875, 3372 and 1160. Her two older children had survived and her husband remarried in 1877.
[22] Death certificate, 1875, 2407. The baby died at an address in Lonsdale Street in the City.
[23] Death certificate, 1875, 3535, died 27 February, at Exploration Lane, Gipps Ward, pneumonia (one week), mother, Caroline Hildebrand, 18 years, born Victora, para 1 (no birth certificate found), was literate.
[24] Death certificates, 1875, 3118, James Handton/Handlon/Hanlon, son of Ellen, 24, born Ireland, para 1 and Peter Byrnes, labourer, illiterate. They soon married and Ellen had live births in 1876, 1877, 1879, 1880, 1882, 1884, 1886, and twins in 1887. She lost a daughter at six years and the twins died at three months and ten months.
[25] *AMJ*, November 1864, pp. 332–333.
[26] Ibid., p. 343.
[27] Ibid., p. 344.
[28] V. Burke and C. Anderson, 'The Relationship of Dietary Lactose to Refractory Diarrhoea in Infants', *Australian Paediatric Journal*, 1965, 1, pp. 147–11. See also J. Walker-Smith and W. A. Walker, 'the Development of Pediatric Gastroenterology: A Historical Overview', Pediatric Research, 2003, Vol. 53, No. 4, pp. 706–715.
[29] *AMJ*, July 1861, p. 171.
[30] Victorian Parliamentary Papers, Royal Commission on Diphtheria, 1872, p. 13.
[31] Victorian Parliamentary Papers, Royal Commission on Sanitary Condition, 1889, pp. 66–67.
[32] [J. William] Mackenna, 'On the Mortality of Children', op. cit., p. 122.
[33] *The Age*, 10 and 11 January 1872.
[34] *The Age*, 22 January 1875; David Dunstan, *Governing the Metropolis*, Melbourne University Press, Melbourne, 1984, pp. 251–263.
[35] *AMJ*, 15 November 1879, p. 549.
[36] Dr James Jamieson, 'Diseases that Should be Prevented', *AMJ*, September and October 1878, p. 298.

Solar energy

[1] Tony McMichael and Basil Hetzel, *The LS Factor*, Penguin, Melbourne, 1987.
[2] C. Parkes, *An Investigation of the Australian Passion for Sport*, <http://people.hws.edu/mitchell/oz/papers/Parker Oz.html>; R. Howell and M. Howell, *A History of Australian Sport*, Shakespeare Head Press, Sydney, 1987.
[3] Film Australia, *Fighting Nature*, Rewind Episode 14, Sydney, 1998.
[4] ESD Working Group on Energy Use, *Final Report*, Australian Government Publishing Service, Canberra, 1991.
[5] I. Lowe, 'Social Aspects of Energy Use', in W. J. Bouma, G. I. Pearman and M. R. Manning (eds), *Greenhouse: Coping with Climate Change*, CSIRO Publishing, Melbourne, 1996, pp. 450–463.
[6] P. Chapman, *Fuel's Paradise*, Penguin, Harmondsworth, 1975.
[7] Council of Australian Governments, *National Strategy for Ecologically Sustainable Development*, Australian Government Publishing Service, Canberra, 1992.
[8] United Nations Environment Program, *Global Environmental Outlook 2000*, Earthscan, London, 1999.
[9] Ibid.
[10] ESD Working Group on Energy Use, *Final Report*, op. cit.
[11] B. Fleay, *The End of the Age of Oil*, Pluto Press, London, 1997.
[12] I. Lowe, 'Towards Sustainable Energy Systems', in S. Dovers (ed.), *Sustainable Energy Systems*, Cambridge University Press, Cambridge, 1994, pp. 194–216.
[13] United Nations Environment Program, *Global Environmental Outlook 2000*, op. cit.
[14] I. Lowe, *Living in the Greenhouse*, Scribe, Victoria, 1989.
[15] Inter-governmental Panel on Climate Change, *Third Assessment Report*, IPCC, Geneva, 2001.
[16] Ibid.
[17] Ibid.
[18] Inter-governmental Panel on Climate Change, *Second Assessment Report*, IPCC, Geneva, 1996.
[19] Kyoto Protocol to the Framework Convention on Climate Change, United Nations Environment Program, Nairobi, 1997.
[20] Parliament of the Commonwealth of Australia, *The Heat is On*, Report of the Senate Environment, Communications, Information Technology and the Arts References Committee, Commonwealth of Australia, 2000, p. 66.
[21] Australian Greenhouse Office, *National Greenhouse Gas Inventory — Land Use Change and Forestry Sector 1990–1999*, Commonwealth of Australia, Canberra, 2000.
[22] K. Davidson, 'Australia, the Kyoto Shag', *The Age*, 5 September 2002.
[23] National Centre for Economic and Industrial Research, *Subsidies of Natural Resource Industries*, Report to Department of Environment, Sport and Territories, Australian Government Publishing Service, Canberra, 1996.
[24] Australian Greenhouse Office, <http://www.greenhouse.gov.au/energyefficiency/appliances/meps/index.html>.
[25] Organisation for Economic Cooperation and Development, *Eco-Efficiency: From Principle to Practice*,

Proceedings of workshop hosted by the Government of Australia in cooperation with United Nations Environment Program and the World Business Council for Sustainable Development, Geneva, 1999.

[26] R. Watson, Statement to COP6 (Sixth Session of the United Nations Framework Convention on Climate Change (UNFCCC) *Conference of the Parties (COP6)*), The Hague, the Netherlands, 13–24 November 2000), <http://www.unep.ch/ipcc/press/sp-cop6.htm>.

[27] Ibid.

[28] Australian Greenhouse Office, <http://www.greenhouse.gov.au/markets/mret/index.html>.

[29] Energy Market Review, *Final Report: Towards a Truly National and Efficient Energy Market*, Council of Australian Governments, Canberra, 2002.

[30] I. Lowe, *Living in the Greenhouse*, op. cit.

[31] H. Turton, *The Aluminium Smelting Industry: Structure, Market Power, Subsidies and Greenhouse Gas Emissions*, Australia Institute, Canberra, 2002.

[32] H. Saddler, 'Using Energy Efficiently in Buildings and Industry', in S. Dovers (ed.), *Sustainable Energy Systems*, op. cit., pp. 50–75.

[33] Parliament of the Commonwealth of Australia, *The Heat is On*, op. cit.

[34] ESD Working Group on Energy Use, *Final Report*, op. cit.

[35] I. Lowe, D. E. Backhouse and M. Sheumack, 'The Experience of Solar Hot Water Systems', *Search*, Vol. 15, 1984, pp. 165–167.

[36] P. C. Stern and E. Aronson (eds), *Energy Use: The Human Dimension*, W. H. Freeman, New York, 1984.

[37] M. Diesendorf, 'Wind, Biomass and other Renewables', in S. Dovers (ed.), *Sustainable Energy Systems*, op. cit., pp. 150–169.

[38] Enviromission Ltd, <http://www.enviromission.com.au/index1.htm>.

[39] Parliament of the Commonwealth of Australia, *The Heat is On*, op. cit.

[40] M. Green, 'Third Generation Photovoltaics', Paper to Boden Research Conference on Artificial Photosynthesis, Sydney, 2003.

[41] P. C. Stern and E. Aronson (eds), *Energy Use*, op. cit.

[42] I. Lowe, D. E. Backhouse and M. Sheumack, 'The Experience of Solar Hot Water Systems', op. cit.

[43] I. Lowe, 'Towards Sustainable Energy Systems', op. cit., p. 194.

[44] Clive Hamilton, *Running from the Storm*, UNSW Press, Sydney, 2002.

[45] P. Kennedy, *Preparing for the Twenty-First Century*, Harper Collins, London, 1993, p. 345.

[46] I. Lowe, 'Non-Renewable Support', *New Scientist*, No. 2317, 11 January 2003, p. 47.

[47] I. Lowe, 'Who Benefits from Australian Energy Research?', *Science and Public Policy*, Vol. 10, 1983, pp. 278–284; I. Lowe, 'Social and Political Issues', Paper to Boden Research Conference on Artificial Photosynthesis, Sydney, 2003.

[48] P. Raskin et al., *Great Transition: The Promise and Lure of the Times Ahead*, Stockholm Environment Institute, Boston, 2002.

[49] I. Lowe, 'Performance Measurement', *Proceedings of the Fenner Conference on the Environment*, Australian Academy of Science, 1994.

[50] Australian Bureau of Statistics, *Measuring Australia's Progress*, AGPS, Canberra, 2002.

[51] I. Lowe, 'Performance Measurement', op. cit.

[52] World Commission on Environment and Development, *Our Common Future*, Oxford University Press, Oxford, 1987.

Revolutionary weather

[1] N. Nicholls, 'Historical El Niño/Southern Oscillation Variability in the Australasian Region', in H. F. Diaz and V. Markgref, *El Niño: Historical and Paleoclimatic Aspects of the Southern Oscillation*, Cambridge University Press, Cambridge, 1992. See also R. H. Grove, 'The East India Company, the Australians and the El Niño', in R. H. Grove, *Ecology, Climate and Empire: Colonialism and Global Environmental History 1400–1940*, White Horse Press, Cambridge, 1997, pp. 124–146.

[2] R. H. Grove, 'The Global Impact of the 1789–1793 El Niño', Nature, London, 28 May 1998, pp. 14–15. See statement by United States Vice-President Al Gore and NOAA scientists, The White House, 8 June 1998, claiming the 1997–1998 El Niño to be the most significant climatic event of the century, and suggesting (without any evidence being produced) that this implied an acceleration of global warming. Other examples include: Kevin Trenberth and T. J. Hoar, 'The 1990–1995 El Niño-Southern Oscillation Event: Longest on Record?', *Geophysical Research Letters*, Vol. 23, 1996, pp. 57–60.

[3] Governor Phillip's diary, reported in T. McCormick, *First Views of Australia, 1788–1825: An Early History of Sydney*, Sydney, 1987.

[4] Quoted in Neville Nicholls, 'More on Early ENSOs: Evidence from Australian Documentary Sources', *Bulletin of the American Meteorological Society*, Vol. 69, 1988, pp. 4–7.

[5] Alexander Beatson, *Tracts relative to the Island of St Helena*, W. Bulmer and Co., London, 1816, p. 10.

[6] J. Kington, *The Weather of the 1780s over Europe*, Cambridge University Press, Cambridge, 1988.

[7] See Census of Tanjore, Papers of the Walkers of Bowland, MS 13615 B, National Library of Scotland, Edinburgh.

[8] William Roxburgh, Letter to Sir Charles Oakley, dated 23 January 1793, East India Company Boards Collections, Ref. no. F/4/99, p. 29; British Library, India Office Library and Records, [IOLR] London.

[9] William Roxburgh, 'A Meteorological Diary Kept at Fort St. George in the East Indies', *Philosophical Transactions of the Royal Society*, Vol. 68, 1778, pp. 180–190; and Vol. 80, 1790, pp. 112–114.

[10] W. Roxburgh, 'Report to the President's Council, Privy Council Letters, Vol. clxxxl, Tamil Nadu State Archives, dated 8 February 1793.

[11] Manuscript reports on remissions of Company land revenues due to prolonged droughts and famine in Northern Circars districts of the Madras Presidency, East India Company Boards Collections, ref. no. F/4/12-12-14, sections 735–743, British Library, IOLR, London.

[12] Report in the *Madras Courier*, 3 November 1791.

[13] Figures quoted in Simon Commander, 'The Mechanics of Demographic and Economic Growth in Uttar Pradesh: 1800–1900', in Tim Dyson (ed.), *India's Historical Demograph: Studies in Famine, Disease and Society*, Curzon Press, London, 1989, p. 50.

[14] For an overview, see B. Murton, 'Spatial and Temporal Patterns of Famine in Southern India before the Famine Codes', in B. Currey and Graeme Hugo (eds), *Famine as a Geographical Phenomenon*, Reidel Publishing Company, Dordrecht, 1984, pp. 71–89. See also Roland Lardinois, 'Deserted Villages and Depopulation in Rural Tamil Nadu, c. 1780–1830', in T. Dyson (ed.), *India's Historical Demograph*, op. cit., pp. 16–48.

[15] W. H. Quinn et al., 'Historical Trends and Statistics of the Southern Oscillation, El Niño and Indonesian Droughts', *Fishery Bulletin*, Vol. 76, 1975, iii, pp. 663–678.

[16] Personal communication, Thierry Correge, ORSTOM,

Noumea, New Caledonia.

[17] *Log of Captain William Bligh*, 6 December 1788, HM Admiralty Records PRO London; Facsimile portion of log published by Pagemaster Press, Guildford, England, 1981.

[18] *Log of Captain William Bligh*, 18 June 1789.

[19] G. H. Endfield and S. L. O'Hara, 'Conflicts over Water in the "Little Drought Age" in Central Mexico', *Environment and History*, Vol. 3, 1997, pp. 255–272.

[20] Arij Ouweneel, *Shadows over Anahuac: An Ecological Interpretation of Crisis and Development in Central Mexico, 1730–1810*, University of New Mexico Press, Albuquerque, 1996, p. 92.

[21] Ibid., pp. 75–91.

[22] Further ENSO-related crop failures also took place in the summers of 1808–1811, bringing about a wholesale restructuring of the economy of Central Mexico.

[23] Vere Langford Oliver, *The History of the Island of Antigua ... From the First Settlement in 1635 to the Present Time*, 3 volumes, Mitchell and Hughes, London, 1894–1899 (first published 1844), pp. 189–191.

[24] Petition dated 13 August 1791 by William McKealy on behalf of the Council of Montserrat, Leeward Islands, British West Indies; Montserrat Legislative Assembly Proceedings, Government Archives, Plymouth, Colony of Montserrat.

[25] As stated in Letter of 6 March 1792, from the Commissioner of the Council of Trade and Plantations to the Council of Monserrat, Government Archives, Plymouth, Montserrat.

[26] Letter from the Directors of the East India Company to the Governor of St Helena, dated 7 March 1794, St Helena Records, Government Archives, St Helena, South Atlantic, reproduced in H. R. Janisch (ed.), *Extracts of the St Helena Chronicles and Records of Cape Commanders*, Jamestown, St Helena, 1908.

[27] Personal communication, Michael Chenoweth, 1998, who referred to the *Times* Index of 1796.

[28] W. H. Quinn and V. T. Neal, 'El Niño Occurrences over the Past Four and a Half Centuries', *Journal of Geophysical Research*, Vol. 92, 1987, pp. 14449–14461.

[29] Antoune Zakry, *The Nile in the Times of the Pharoahs and the Arabs* (in Arabic), Maktabet Al-Ma'arif, Cairo, 1926 (reprinted 1995 by Maktabet Madbouli).

[30] W. Willocks and J. I. Craig, *Egyptian Irrigation*, 3rd edition, Vols 1 and 2, E. and F. N. Spon, London, 1913.

[31] C. Webb and J. B. Wright, *The James Stuart Archive of Recorded Oral Evidence Relating to the History of the Zulu and Neighbouring Peoples*, Vol. 1, University of Natal Press, Pietermaritzburg, South Africa, 1976.

[32] The 1862 event caused the worst droughts in southern Africa in the nineteenth century. Dendrochronological records point to the early 1860s and the 1790s as long dry spells.

[33] M. Hall, 'Dendrochronology, Rainfall and Human Adaptation in the Later Iron Age of Natal and Zululand', *Annals of the Natal Museum*, 1976, p. 702.

[34] *The Australian*, 24 December 1998 (note that similar unusually heavy ice conditions caused the wreck of Shackleton's ship *Endurance* in October 1914).

[35] See J. Neumann, 'Great Historical Events that were Significantly Affected by the Weather: Part 2. The Year Leading to the Revolution of 1789 in France', *Bulletin of the American Meteorological Society*, Vol. 58, 1977, pp. 163–168; and J. Neumann and J. Dettwiller, 'Great Historical Events that were Significantly Affected by the Weather: Part 9. The Year Leading to the Revolution of 1789 in France (11)', *Bulletin of the American Meteorological Society*, Vol. 71, 1980, pp. 33–41.

[36] E. L. Ladurie, *Times of Feast, Times of Famine*, Allen and Unwin, London, 1972, p. 75; J. M. Desbordes, *La Chronique Villageoise de Vareddes (Seine et Marne) aux XVIIe et XVIIIe Siecles*, Paris, (edition de l'Ecole, 11 Rue de Sevres) about 1969.

[37] G. Lefebvre, *La Grande Peur de 1789*, Société d'Edition d'Enseignement Superieur, Paris, 1932, pp. 47–53.

[38] In this year the Bordeaux wine was said to have frozen in Louis XIV's goblet. See W. Gregory Monahan, *Year of Sorrow: The Great Famine of 1709 in Lyon*, Ohio State University Press, Columbus, 1993.

[39] See Arthur Young, *Travels in France*, C. Maxwell (ed.), Cambridge University Press, London, 1953, p. 173.

[40] Quoted in Simon Schama, *Citizens: A Chronicle of the French Revolution*, Penguin, London, 1989, p. 331.

[41] Alexander de Tocqueville, *L'Ancien Regime et la Revolution*, J. P. Mayer (ed.), Gallimard, Paris, 1952, p. 96.

[42] M. Barriendos, J. C. Pena, J. Martin-Vide, P. Jonsson and G. Demarée, 'The Winter of 1788–1789 in the Iberian Peninsula from Meteorological Readings, Observations and Proxy-Data Records', in *Giuseppe Toaldo and his Time*, Bertoncello artigrafiche, Padova, 2000.

[43] F. Salva, Archive of the Royal Academy of Medicine, Barcelona, *Tablas Meteorologicas*, Vol. 1, 1788.

[44] J. Fontana, *Historia de Catalunya: La Fie de l'Antic Regim i la Industrialitzacio* (1787–1868), Edicions 62, Vol. 5, Barcelona, quoted in M. Barriendos et al., 'The Winter of 1788–1789 in the Iberian Peninsula', op. cit.

[45] J. Mercader, *Els capitans generals*, Viviens-Vives, Barcelona, 1986; S. Riera i Tuebols, *Ciencia i Tecnic a la Illistracio: Francesc Salva i Campillo (1751–1828)*, La Magrana, Barcelona, 1985; J. M. Torras i Ribe, 'La Catalunya Borbonica: Evolucio i Reaccions contra el Nou Regim', *Historia de Catalunya*, Vol. 4, 1978, pp. 178–203 (Salvat, Barcelona), all quoted in ibid.

[46] M. Barriendos et al., 'The Winter of 1788–1789 in the Iberian Peninsula', op. cit.

[47] *A Country Parson: James Woodforde's Diary 1758–1802*, Oxford University Press, Oxford, 1985. Temperatures were comparable to the high temperatures of January 1998.

[48] Alexander de Tocqueville, *L'Ancien Regime et la Revolution*, op. cit., p. 249. For a more developed discussion of the agrarian background to the French Revolution, see Alan Davies, 'The Origins of the French Revolution of 1789', *History*, Vol. 49, 1964, pp. 24–41.

[49] I am indebted for archival details of the weather in the Great Plains in 1790–1793 to Ted Binnema. See also Theodore Binnema, *Common and Contested Ground: A Human and Environmental History of the North-western Plains*, University of Oklahoma Press, Norman, 2001.

[50] K. R. Foster, M. F. Jenkins and A. C. Toogood, 'The Philadelphia Yellow Fever Epidemic of 1793', *Scientific American*, No. 279, 1998, pp. 68–74.

[51] T. J. Pettigrew, 'Memoirs of the Life and Writings of J. Coakley Pettigrew', Vol. 3, p. 234, quoted in T. Thompson, *Annals of Influenza or Epidemic Catarrhal Fever from 1510 to 1837*, Sydenham Society, London, 1852, pp. 199–202.

[52] Ibid.

[53] See Neville Nicholls, Chapter 2, this volume.

[54] See also R. H. Grove, 'The East India Company, the Australians and the El Niño', op. cit., p. 145.

[55] R. H. Grove, 'Global Impact', op. cit.

Climate change and human health

[1] P. M. Vitousek, H. A. Mooney and J. Lubchenco, 'Human Domination of Earth's Ecosystems', *Science*,

No. 277, 1997, pp. 494–499.

[2] M. Wackernagel, N. B. Schulz, D. Deumling, et al., 'Tracking the Ecological Overshoot of the Human Economy', PNAS Early Edition, <http://www.pnas.org/cgi/doi/10.1073/pnas.142033699>, 27 June 2002.

[3] Intergovernmental Panel on Climate Change (IPCC), *Climate Change 2001, Third Assessment Report* (Working Group II: Impacts and Adaptations), Cambridge University Press, Cambridge, 2001.

[4] E. Lindgren and R. Gustafson, 'Tick-borne Encephalitis in Sweden and Climate Change', *Lancet*, Vol. 358, 2001, pp. 16–18.

[5] J. A. Patz et al., 'Increasing Incidence of Malaria since 1970 Parallels Regional Warming in East Africa', *Nature*, Vol. 420, 2002, pp. 627–628.

[6] S. I. Hay, J. Cox, D. J. Rogers et al., 'Climate Change and the Resurgence of Malaria in the East African Highlands', *Nature*, Vol. 415, 2002, pp. 905–909.

[7] X. Rodo, M. Pascual, G. Fuchs et al., 'ENSO and Cholera: A Non-Stationary Link Related to Climate Change?', *Proceedings, National Academy of Science*, Early Edition 2002; <www.pnas.org/cgi/doi/10.1073/pnas.182203999>. But see also Chapter 11, on ENSO in India.

[8] Intergovernmental Panel on Climate Change, *Climate Change 2001*, op. cit. Also, Department of Health (UK), *Health Effects of Climate Change in the UK*, Department of Health, London, 2002.

[9] R. Woodruff, C. Guest, M. G. Garner et al., 'Predicting Ross River Virus Epidemics from Regional Weather Data', *Epidemiology*, Vol. 13, 2002, p. 384; S. Tong, P. Bi, K. Donald et al., 'Climate Variability and Ross River Virus Transmission', *Journal of Epidemiology and Community Health*, Vol. 56, 2002, pp. 617–621.

[10] G. V. Hall, R. De Souza and M. D. Kirk, 'Foodborne Disease in the New Millenia: Out of the Frying Pan and into the Fire?', *Medical Journal of Australia*, Vol. 177, Nos. 11–12, 2002, pp. 614–618.

[11] D. L. Albritton, M. R. Allen, A. P. M. Baede et al., *Climate Change 2001: The Scientific Basis*, Intergovernmental Panel on Climate Change (IPCC) Working Group I: Summary for Policy Makers, Third Assessment Report, New York, Cambridge University Press, 2001.

[12] Hippocrates, 'Airs, Waters and Places'. An essay on the influence of climate, water supply and situation on health in G. E. R. Lloyd (ed.), *Hippocratic Writings*, Penguin, London, 1978, p. 148.

[13] W. H. McNeill, *Plagues and People*, Penguin, Middlesex, 1976.

[14] H. Pringle et al., 'Death in Norse Greenland', *Science*, No. 275, 1997, pp. 924–926; A. J. McMichael, *Human Frontiers, Environments and Disease: Past Patterns, Uncertain Futures*, Cambridge University Press, Cambridge, 2001.

[15] R. I. Rotberg and T. K. Raab, *Hunger and History*, Cambridge University Press, Cambridge, 1985; R. E. Bryson and T. J. Murray, *Climates of Hunger: Mankind and the World's Changing Weather*, University of Wisconsin Press, Madison, 1977.

[16] B. Fagan, *Floods, Famines and Emperors: El Niño and the Fate of Civilisations*, Basic Books, New York, 1999.

[17] R. H. Grove, 'Revolutionary Weather: The Climatic and Economic Crisis of 1788–1795 and the Discovery of El Niño', (this volume, Chapter 11).

[18] J. S. Duchin, F. T. Koster, C. J. Peters et al., 'Hantavirus Pulmonary Syndrome: A Clinical Description of 17 Patients with a Newly Recognized Disease', *New England Journal of Medicine*, Vol. 330, 1994, pp. 949–955.

[19] P. Wilkinson, A. McMichael, S. Kovats et al., 'International Study of Temperature and Heatwaves on Urban Mortality in Low-and-Middle-Income Countries' (ISOTHURM), Presented at International Society of Environmental Epidemiology, Vancouver, August 2002, *Epidemiology*, Vol. 13(4):S81.

[20] IPCC, *Climate Change 2001*, op. cit.

[21] P. C. D. Milly, R. T. Wetherald, K. A. Dunne et al., 'Increasing Risk of Great Floods in a Changing Climate', *Nature*, Vol. 415, 2002, pp. 514–517.

[22] IPCC, *Climate Change 2001*, op. cit.

[23] W. J. M. Martens, R. S. Kovats, S. Nijhof et al., 'Climate Change and Future Populations at Risk of Malaria', *Global Environmental Change*, Vol. 9 Supplement, 1999, pp. S89–S107.

[24] W. J. M. Martens, J. Rotmans and D. S. Rothman, 'Integrated Assessment Modelling of Human Health Impacts', in W. J. M. Martens and A. J. McMichael (eds), *Environmental Change, Climate and Health: Concepts and Research Methods*, Cambridge University Press, Cambridge, 2002, pp. 197–255.

[25] W. Checkley, L. D. Epstein, R. H. Gilman et al., 'Effects of El Niño and Ambient Temperature on Hospital Admissions for Diarrhoeal Diseases in Peruvian Children', *Lancet*, 355 (9202), 2000, pp. 442–450; R. B. Singh, S. Hales, N. de Wet et al., 'The Influence of Climate Variation and Change on Diarrheal Disease in the Pacific Islands', *Environment Health Perspective*, Vol. 109 (2), 2001, pp. 55–59.

[26] M. L. Parry, C. Rosenzweig, A. Iglesias, G. Fischer and M. Livermore, 'Climate Change and World Food Security: A New Assessment', *Global Environmental Change*, Vol. 9, 1999, S51–S67.

[27] M. L. Parry and T. Carter, *Climate Impact and Adaptation Assessment*, Earthscan, London, 1998.

[28] A. J. Woodward et al., 'Protecting Human Health in a Changing World: The Role of Social and Economic Development', *Bulletin of the World Health Organisation*, Vol. 78, 2000, pp. 1148–1155.

[29] A. J. McMichael, 'The Biosphere, Health and "Sustainability" ', *Science*, No. 297, 2002, p. 1093.

[30] E. O. Wilson, *Consilience: On the Unity of Knowledge*, Alfred Knopf, New York, 1998.

The roaring forties

[1] Jan DeBlieu, *Wind: How the Flow of Air Has Shaped Life, Myth, and the Land*, Houghton Mifflin Company, Boston and New York, 1999, p. 3.

[2] W. S. Jevons, 'Meteorological Observations in Australia', *The Sydney Magazine of Science and Art*, Vol. 2, 1859, p. 176.

[3] Henry A. Hunt, 'An Essay on Southerly Busters', in Ralph Abercromby (ed.), *Three Essays on Australian Weather*, Frederick W. White, Sydney, 1896, pp. 16–62.

[4] Genevieve Miller, ' "Airs, Waters, and Places" in History', *Journal of the History of Medicine*, 17 January 1962, pp. 129–140.

[5] Warwick Anderson, *The Cultivation of Whiteness: Science, Health and Racial Destiny in Australia*, Melbourne University Press, Melbourne, 2002, p. 31.

[6] Ibid., chapter 1. See, for example, P. Divorty, 'The Influence of an Australian Climate on the Constitution of the Western European', *Sydney Magazine of Science and Art*, Vol. 1, 1858, pp. 83–91: 'the Australian climate will in the long run wear out the human constitution' (p. 90).

[7] J. M. Powell, 'Medical Promotion and the Consumptive Immigrant to Australia', *The Geographical Review*, Vol. 63, No. 4, October 1973, pp. 449–476; *Mirrors of the New World*, Archon Books, Kent, 1977; Pat Jalland,

Australian Ways of Death, Oxford University Press, Melbourne, 2002, pp. 40–45.

[8] J. W. Barrett, *Victoria and Tasmania as Resorts for Consumptives and Persons Afflicted with Lung Disease*, Harrison and Sons, London, 1884 (italics in original).

[9] Jean-René Vanney, *Histoire des Mers Australes*, Fayard, Paris, 1986, p. 22.

[10] W. L. N. Tickell, *Albatrosses*, Yale University Press, New Haven and London, 2000, p. 261.

[11] See Glyndwr Williams, *The Great South Sea: English Voyagers and Encounters 1570–1750*, Yale University Press, New Haven and London, 1997.

[12] E. A. Wilson, *Aves*, Vol. 2, 1907, quoted in William Jameson, *The Wandering Albatross*, Rupert Hart-Davis, London, 1958, p. 23.

[13] William Jameson, *The Wandering Albatross*, op. cit. and W. L. N. Tickell, *Albatrosses*, op. cit.

[14] J. H. Parry, *The Discovery of the Sea*, University of California Press, Berkeley, 1981, chapter XII.

[15] Ibid., p. 86.

[16] Jan DeBlieu, *Wind*, op. cit., p. 50.

[17] See Frank Broeze, *Island Nation: A History of Australians and the Sea*, Allen and Unwin, Sydney, 1998.

[18] Geoffrey Blainey, *The Tyranny of Distance*, Sun Books, Melbourne, 1966.

[19] Ross Gibson, 'Ocean Settlement', *Meanjin*, Vol. 53, No. 4, Summer 1994, pp. 665–678, at pp. 666–667.

[20] Grace Karskens, *The Rocks*, Melbourne University Press, Melbourne, 1997, p. 18 and p. 183.

[21] Edward Snell, *The Life and Adventures of Edward Snell*, ed. Tom Griffiths and Alan Platt, Angus and Robertson, Sydney, 1988, pp. 25–26 (October 9, 1849).

[22] Ibid., pp. 27–28 (October 13, 1849).

[23] Ibid., p. 384 (April 10, 1858).

[24] Ibid., p. 386 (April 27, 1858).

[25] Geoffrey Blainey, *The Tyranny of Distance*, op. cit., p. 177.

[26] Matthew Fontaine Maury, *The Physical Geography of the Sea and its Meteorology*, ed. John Leighly, the Belknap Press of Harvard University Press, Cambridge, 1963, [eighth edition, originally published 1861], p. 307.

[27] Ibid., p. 306.

[28] Alan Platt, 'The City of Adelaide', unpublished manuscript on the economic and social history of the tall ships, 2002, in private possession.

[29] Les Murray, *The Australian Year*, Angus and Robertson, Sydney, 1985, p. 13.

[30] Alan Frost, *Botany Bay Mirages*, Melbourne University Press, Melbourne, 1994, chapter 7.

[31] Quoted in ibid., p. 156.

[32] R. A. Swan, *Of Myths and Mariners*, R. A. Swan, Sydney, no date, copy in Scott Polar Research Institute, Cambridge.

[33] David Campbell, *The Crystal Desert: Summers in Antarctica*, Houghton Mifflin Company, Boston, New York, London, 1992, chapter 8.

[34] Jean M. Grove, *The Little Ice Age*, Methuen, London and New York, 1988; ibid., p. 161.

[35] Max Downes, 'First Visitors to Heard Island', *ANARE Research Notes No. 104*, no date [1990s], Australian Antarctic Division, Hobart.

[36] John Gardner, *Stormy Weather: A History of Research in the Bureau of Meteorology*, Metarch Papers No. 11, December 1997, Bureau of Meteorology, Melbourne, 1997.

[37] H. C. Russell, 'Moving Anticyclones in the Southern Hemisphere', in R. Abercromby (ed.), *Three Essays on Australian Weather*, op. cit., pp. 1–15.

[38] Griffith Taylor, 'Climatic Relations between Antarctica and Australia', in W. L. G. Joerg (ed.), *Problems of Polar Research*, American Geographical Society, Special Publication no. 7, New York, 1928, pp. 285–300.

[39] M. F. Maury to Robert FitzRoy, 25 February 1859, quoted in W. J. Gibbs, *The Origins of Australian Meteorology*, Australian Government Publishing Service, Canberra, 1975, p. 9.

[40] Robert Mare Friedman, *Appropriating the Weather: Vilhelm Bjerknes and the Construction of a Modern Meteorology*, Cornell University Press, Ithaca and London, 1989.

[41] For example, J. W. Gregory, *The Climate of Australasia: In Reference to its Control by the Southern Ocean*, Whitcombe and Tombs, Melbourne, 1904.

[42] C. T. Madigan, 'Tabulated and Reduced Records of the Cape Denison Station, Adelie Land', in Australasian Antarctic Expedition 1911–1914, *Scientific Reports*, Series B, Vol. 4, Meteorology, Alfred J. Kent, Government Printer, Sydney, 1929, p. 20.

[43] Douglas Mawson, 'Preface to Meteorological Publications', *Scientific Reports*, Series B, Vol. 3, p. 5.

[44] Morton Henry Moyes, Diary, Mitchell Library, Sydney, ML MSS 388/1 (CY 3660).

[45] Ibid., 1912.

[46] *Herald* (Melbourne), 22 December 1915.

[47] *The Dominion*, 27 August 1913.

[48] G. Ainsworth to H. Hunt, 12 August 1913, MS, Bureau of Meteorology, kindly brought to my attention by Neville Nicholls. See also G. Ainsworth's chapters in Douglas Mawson, *The Home of the Blizzard*, William Heinemann, London, 1915.

[49] M. F. Maury, *Physical Geography of the Sea*, op. cit., pp. 406–415.

[50] Contrary to Maury's theory of a milder south pole, Cook's naturalist on his second voyage, Johann Reinhold Forster, had earlier concluded that southern high latitudes were consistently colder than northern ones. J. R. Forster, *Observations Made During a Voyage Around the World*, (ed. N. Thomas, H. Guest and M. Dettelbach), University of Hawai'i Press, Honolulu, 1996, p. 37.

[51] Douglas Mawson, 'The Unveiling of Antarctica', [ANZAAS Presidential Address], in G. W. Leeper (ed.) *Report of the Twenty-Second Meeting of the Australian and New Zealand Association for the Advancement of Science, Melbourne Meeting, January, 1935*, H. J. Green, Government Printer, Melbourne, 1935, p. 31.

[52] Stephen J. Pyne, *The Ice: A Journey to Antarctica*, University of Iowa Press, Iowa City, 1986, pp. 41–42.

[53] The quote is from ibid., p. 50.

Cyclone Tracy

[1] Voices from ABC Radio National '*That's History: Cyclone Tracy*' (with Bill Bunbury) broadcast, March 1991.

[2] C. D. Haynes, M. G. Ridpath and M. A. J. Williams (eds), *Monsoonal Australia: Landscape, Ecology and Man in the Northern Lowlands*, A. A. Balkema, Rotterdam, 1991; Bureau of Meteorology, *Understanding Cyclones*, Bureau of Meteorology, Canberra, c. 1991.

[3] Bill Bunbury, *Cyclone Tracy: Picking up the Pieces / Talking History with Bill Bunbury*, Fremantle Arts Centre Press, South Fremantle, 1994.

[4] For scholarly reflections on the sensitivity of public memory in Darwin immediately after the cyclone and twenty years later, see Peter Read, *Returning to Nothing: The Meaning of Lost Places*, Cambridge University Press, Cambridge, 1996, chapter 7, and Mickey Dewar, 'Reflections on the Cyclone Tracy Exhibition at the Museum and Art Gallery of the Northern Territory', Unpublished paper presented to the Australian Historical Association conference, 1996.

Palaeoclimates

[1] For examples, see M. Williams, D. Dunkerley, P. De Deckker, P. Kershaw and J. Chappell, *Quaternary Environments*, Arnold, London, 1998 (second edition).

[2] A. Penck and E. Brückner, *Die Alpen im Eiszeitalter*, Tauchnitz, Leipzig, 1909. The idea of an 'Ice-Age' gained currency through the nineteenth century. The poet Goethe spoke of *'an epoch of great cold'* and the geologist Louis Agassiz coined the term *Eiszeit* or Ice Age for what we now call the Pleistocene. For a brief history of these ideas, see Andrew Goudie, *Environmental Change*, Clarendon Press, Oxford, 1977, pp. 4–7.

[3] Cited in G. Daniel, *A Hundred and Fifty Years of Archeology*, G. Duckworth, London, 1975 (second edition), p. 27. Frere was reporting his discovery in 1797 of antediluvian hand axes and the bones of extinct animals in Thames valley sediments at Hoxne.

[4] C. Emiliani, 'Pleistocene Temperatures', *Journal of Geology*, Vol. 63, 1955, pp. 538–578.

[5] Conversation with James Arnold at Australian National University in August 1993. See also J. R. Arnold and W. F. Libby, 'Age Determinations by Radiocarbon Content: Checks with Samples of Known Age', *Science*, No. 110, 1949, pp. 678–680.

[6] C. Emiliani, op. cit. For an overview of the isotopic chemistry involved, see M. J. Aitken, *Science-Based Dating in Archaeology*, Longman, London, 1990.

[7] N. J. Shackleton and D. Opdyke, 'Oxygen Isotope and Palaeomagnetic Stratigraphy of Equatorial Pacific Core V28-238: Oxygen Isotope Temperatures and Ice Volumes on a 10^5 Year and 10^6 Year Scale', *Quaternary Research*, Vol. 3, 1973, pp. 39–55. However, variability in the temperature of the abyssal ocean remains a factor, as J. Chappell and N. J. Shackleton, 'Oxygen Isotopes and Sea Level', *Nature*, 1986, Vol. 324, pp. 137–140, have shown through an analysis of discrepancies between the marine isotope record and sea-level curves derived from dated coral terraces.

[8] See M. J. Aitken, *Science-Based Dating in Archaeology*, Longman, London, 1990. The original stage numbers were allocated by C. Emiliani. As the palaeoclimate record became more detailed, the scheme had to expand to include lettered substages (e.g. 5e). These were later replaced by a decimal system (e.g. stage 5.5) which allowed for even more detail.

[9] D. G. Martinson, N. G. Pisias, J. D. Hays, J. Imbrie, T. C. Moore and N. J. Shackleton, 'Age Dating and the Orbital Theory of the Ice Ages: Development of a High-Resolution 0 to 300,000-year Chronostratigraphy', *Quaternary Research*, Vol. 27, 1987, pp. 1–29.

[10] M. Williams, D. Dunkerley, P. De Deckker, P. Kershaw and J. Chappell, *Quaternary Environments*, Arnold, London, 1998 (second edition), p. 49. These rapid temperature fluctuations are called Dansgaard-Oeschger events after the researchers who first described them. See J. Adams, M. Maslin and Ellen Thomas, 'Sudden Climate Changes during the Quaternary', <http://www.esd.ornl.gov/projects/qen/transit.html>.

[11] John Chappell, 'Climate before Agriculture', in A. Anderson, I. Lilley and S. O'Connor, *Histories of Old Ages: Essays in Honour of Rhys Jones*, Pandanus Books, Canberra, 2001, pp. 171–183, especially p. 173.

[12] Yet another heavy isotope — deuterium, an isotope of hydrogen — provides a proxy measure of temperature during this accumulation.

[13] B. Stauffer, 'Climate Change: Cornucopia of Ice Core Results', *Nature*, 1999, Vol. 399, pp. 412–413 and J. R. Petit, J. Jouzel, D. Raynaud, N. I. Barkov et al., 'Climate and Atmospheric History of the Past 420,000 Years from the Vostok Ice Core, Antarctica', *Nature*, 1999, Vol. 399, pp. 429–436.

[14] M. Williams, D. Dunkerley, P. De Deckker, P. Kershaw and J. Chappell, *Quaternary Environments*, Arnold, London, 1998 (second edition).

[15] J. Adams et al., 'Sudden Climate Changes during the Quaternary', op. cit.

[16] Broecker, paraphrased in ibid.

[17] D. Walker, 'Limitations of Some Biological Bases of Palaeoclimatology', *Australian Meteorological Magazine*, Vol. 24, No. 1, 1976, pp. 21–32. Cited in R. Baker, 'The Aboriginal and Environmental History of the Chambers Bay Coastal Plains', 1981, unpublished BA (Hons) thesis, Australian National University, Canberra, p. 23.

[18] The impact of ENSO has been intellectual as well as climatic, unleashing a new enthusiasm for climatic determinism.

[19] P. Veth, S. O'Connor and L. Wallis, 'Perspectives on Ecological Approaches in *Australian Archaeology', Australian Archaeology (Special 50th Edition — Archaeology in the New Millennium)*, Vol. 50, 2000, pp. 54–66. The authors conclude (p. 61) that 'despite the problems inherent in the use of palaeoecological data and modelling, the archaeological record cannot be profitably divorced from its environmental context'.

[20] M. J. Rowland, 'Holocene Environmental Variability: Have its Impacts Been Underestimated in Australian Pre History?' *The Artefact*, Vol. 22, 1999, pp. 11–40.

[21] Colonial scientists were alert to the possibilities of using changes in sea levels as a marker of geological antiquity, and they made careful records of any field evidence that came their way. One of the earliest records is by Robert Etheridge in 1896 between Redfern and Botany Bay in Sydney. Etheridge, curator of geology and palaeontology at the Australian Museum, noted the stumps of large eucalypts 'at a depth of fifteen feet below high water'. He concluded 'the Pacific Ocean and the Australian land have changed their respective levels by as much as fifteen feet, since the existence of Neolithic man at Botany Bay'.

[22] K. Lambeck, T. M. Esat and E-K Potter, 'Links between Climate and Sea Levels for the Past Three Million Years', *Nature*, Vol. 419, 2002, pp. 199–202.

[23] The apparent rate of sea-level rise will vary from region to region, because of changes in the extent of isostatic deformation of the Earth's surface due to the shifting weight of ice and water. However, M. Williams et al., *Quaternary Environments*, op. cit., p. 120, note that 'stepped deglaciations, and hence changing rates of sea-level rise marine transgression, appear to have been the norm in the late Quaternary'. K. Lambeck and J. Chappell, in 'Sea Level Change through the Last Glacial Cycle', *Science*, No. 292, 2001, pp. 679–686, note two periods of rapid and sustained global sea-level rise of about 15 m per 1,000 years, approximately 16,000 to 12,500 years ago and from 11,500 to 8,000 years ago.

[24] M. A. Smith, M. Spriggs and B. Fankhauser (eds), *Sahul in Review: Pleistocene Archaeology in Australia, New Guinea and Island Melanesia*, Department of Prehistory, Australian National University, Canberra, 1993; J. Allen and P. Kershaw, 'The Pleistocene–Holocene Transition in Greater Australia', in L. G. Straus, B. V. Eriksen, J. M. Erlandson and D. R. Yesner, *Humans at the End of the Ice-Age: The Archaeology of the Pleistocene-Holocene Transition*, Plenum, New York, 1996, pp. 175–199.

[25] Both the shape of the continental shelf and the rate of sea-level rise determine the rate of inundation of coastal areas. For north-western Australia, M. Williams et al., *Quaternary Environments*, op. cit., p. 120, suggest that the shoreline retreated at a rate of 20 m per year (1 m per fortnight). An earlier estimate by T. H. van Andel and J. J. Veevers, *Morphology and Sediments of the Timor Sea*, Bureau of Mineral Resources, Geology and Geophysics, Canberra. Bulletin 83, 1967, p. 109, suggests rates of

about 5.5 km per year (10–15 m per day) across the centre of the Sahul shelf during the early part of the transgression. For the Nullarbor coast, R. V. S. Wright (ed.), *Archaeology of the Gallus Site, Koonalda Cave*, Australian Institute of Aboriginal Studies, Canberra, 1971 p. 13, calculated that between 14,000 and 13,000 years ago, land was lost at a rate of about 50 m per year (1 m per week). On the low gradient continental shelf at Princess Charlotte Bay, north Queensland, J. M. Beaton suggests rates of shoreline displacement of 300 to 400 m per year (1 m per day) or more in 'Evidence for a Coastal Occupation Time Lag at Princess Charlotte Bay (North Queensland) and Implications for Coastal Colonisation and Population Growth Theories for Aboriginal Australia', *Archaeology in Oceania*, Vol. 20, 1985, pp. 1–20, derived from R. W. Galloway and E. Loeffler, 'Aspects of Geomorphology and Soils in the Torres Strait Region', in D. Walker (ed.), *Bridge and Barrier: The Natural and Cultural History of Torres Strait*, Australian National University, Canberra, 1972, pp. 11–28.

[26] The term 'landesque capital' was used — possibly coined — by H. C. Brookfield, 'Intensification Intensified', *Archaeology in Oceania*, Vol. 21, 1986, pp. 177–180, to refer to agricultural infrastructure such as terraces, dams, irrigation systems and improved soil. But the concept can be applied to hunter-gatherer and pastoral landscapes. 'The innumerable stock-dams which dot the Australian landscape are landesque capital', says Brookfield, and 'some landesque capital is so enduring that it can be re-used thousand of years after abandonment, given a new input of labour to get it back into working order'.

[27] Geoffrey Blainey, *The Triumph of the Nomads*, Macmillan, Melbourne, 1975 (first edition), pp. 89–91; John Mulvaney in *The Prehistory of Australia*, Penguin, Melbourne, 1975 (second edition), p. 136, speculated that it would have 'required adaptive efforts and a consolatory philosophy'.

[28] S. Bowdler, 'The Coastal Colonisation of Australia', in J. Allen, J. Golson and R. Jones (eds), *Sunda and Sahul: Prehistoric Studies in Southeast Asia, Melanesia and Australia*, Academic Press, New York, 1977, pp. 205–246, especially p. 223.

[29] J. M. Beaton, 'Evidence for a Coastal Occupation Time Lag', op. cit., drew attention to a time lag of several millennia between the stabilisation of sea level and the earliest archaeological evidence for coastal subsistence economies. He suggested that shell populations requiring calm and shallow waters above sandy or muddy floors could only have developed after sea level stabilised — and would then take several millennia to establish extensive shell-beds.

[30] Details are summarised in S. Cane, 'The Great Flood: Eustatic Change and Cultural Change in Australia during the Late Pleistocene and Holocene', in A. Anderson, I. Lilley and S. O'Connor, *Histories of Old Ages: Essays in Honour of Rhys Jones*, Pandanus Books, Australian National University, 2001, pp. 141–165.

[31] Ibid., p. 147.

[32] For a review, see Sandra Bowdler, 'Offshore Islands and Maritime Explorations in Australian Prehistory', *Antiquity*, Vol. 69, 1995, pp. 945–958. Although some archaeologists, like Charlie Dortch (C. E. Dortch, 'Preliminary Underwater Survey for Rock Engravings and other Sea Floor Sites in the Dampier Archipelago, Pilbara Region, Western Australia', *Australian Archaeology*, Vol. 54, 2002, pp. 37–42), have surveyed the sea floor for drowned sites (without success), the best archaeological evidence has come from caves and rock shelters that were once inland sites on the continental shelves.

[33] Sandra Bowdler, *Hunter Hill, Hunter Island: Archaeological Investigations of a Prehistoric Tasmanian Site*, Australian National University, Canberra *(Terra Australis 8)*, 1984, p. 138.

[34] Ibid. People did not return to Hunter Island until 2,500 years ago, when Aboriginal people in this region began to use water-craft to visit off-shore islands.

[35] Sandra Bowdler, 'Offshore Islands and Maritime Explorations in Australian Prehistory', op. cit., p. 946.

[36] C. Woodroffe, B. G. Thom and J. Chappell, 'Development of Widespread Mangrove Swamps in Mid-Holocene Times in Northern Australia', *Nature*, Vol. 317, 1985, pp. 711–713.

[37] For a review of Arnhem Land rock art and prehistory, see P. S. C. Taçon and S. Brockwell, 'Arnhem Land Prehistory in Landscape, Stone and Paint', *Antiquity*, Vol. 69, 1995, pp. 676–695.

[38] For example, see M. A. Smith, 'Radiocarbon Dates for Bifacial Points at Scotch Creek 1, Northern Territory', *Australian Archaeology*, Vol. 41, 1995, pp. 40–41; and M. A. Smith and C. J. Brockwell, 'Archaeological Investigations at a Stratified Open Site near Humpty Doo, Northern Australia', in M. Sullivan, S. Brockwell and A. Webb (eds), *Archaeology in the North: Proceedings of the 1993 Australian Archaeological Association Conference*, North Australia Research Unit, Australian National University, Darwin, 1994, p. 84. Stone points are associated in the ethnography for northern Australia with light-weight duelling spears. The very large number of points on some sites on the coastal plains, and their predominance in assemblages at this time, suggest a period of concerted point manufacture. This in turn suggests greater inter-group conflict starting approximately 3,500 years ago, either directly as in the case of warfare, or indirectly through increased threat and display at ceremonies.

[39] P. S. C. Taçon and C. Chippindale, 'Australia's Ancient Warriors: Changing Depictions of Fighting in the Rock Art of Arnhem Land, N. T.', *Cambridge Archaeological Journal*, Vol. 4, 1994, pp. 211–214.

[40] In *Civilizations* (Macmillan, London, 2000), Felipe Fernández-Armesto calls grasslands 'the sweepings of the winds'. I apply the term to deserts because aeolian landforms are striking features of desert landscapes.

[41] J. N. Jennings, 'A Revised Map of the Desert Dunes of Australia', *Australian Geographer*, Vol. 10, 1968, pp. 408–409; and J. N. Jennings, 'Desert Dunes and Estuarine Fill in the Fitzroy Estuary (North-Western Australia)', *Catena*, Vol. 2, 1975, pp. 215–216.

[42] For discussion of the behavioural responses of human groups in the Australian desert to changes in glacial-age climates — and also to more subtle climatic oscillations in the Holocene — see P. D. Hiscock, 'Prehistoric Settlement Patterns and Artefact Manufacture at Lawn Hill, North West Queensland', unpublished PhD thesis, University of Queensland, Brisbane, 1988; M. A. Smith, 'The Case for a Resident Human Population in the Central Australian Ranges during Full Glacial Aridity', *Archaeology in Oceania*, Vol. 24, 1989, p. 93; M. A. Smith, 'Biogeography, Human Ecology and Prehistory in the Sandridge Deserts', in T. Murray (ed.), *Archaeology of Aboriginal Australia: A Reader*, Allen and Unwin, Melbourne, 1998, p. 119; and P. Veth, 'Aridity and Settlement in North West Australia', *Antiquity*, Vol. 69, 1995, p. 733.

[43] R. A. Gould, *Puntutjarpa Rockshelter and the Australian Desert Culture*, American Museum of Natural History, New York, 1977.

[44] Ibid., p. 182.

[45] The earliest attempt to read a history into central Australian landscapes was by the explorer Charles Sturt in 1845. After dragging a whaleboat deep into the arid interior in the expectation of finding an inland sea, Sturt's expedition was blocked by the sand ridges of the Simpson

Desert. Sturt thought this dune field had a marine origin: *'They exhibit a regularity that water alone could have given, and to water, I believe, they plainly owe their first existence'.* Because of the difficulty of imagining the contemporary processes that actively shape these landscapes, early European observers often regarded deserts as relict marine forms. Sturt was wrong about the processes involved, but correct in interpreting these massive sand ridge deserts as relict landforms. The Australian arid zone is ringed by fossil dunes aligned with the dominant dry season winds of the late Pleistocene. More concrete evidence of environmental and climatic change in the interior was the remarkable discovery in 1893 of the fossil remains of hundreds of extinct giant marsupials and ratite birds, on the now dry bed of Lake Callabonna in Central Australia. A Museum field party, organised by E. C. Stirling, director of the South Australian Museum, excavated numerous remains of the giant marsupial *Diprotodon*, found mired in the blue and grey muds of the salt lake. The finds created a powerful popular image of a better watered, more fertile ancient Australia preceding the present aridity of these areas.

[46] Palaeoenvironmental conditions are summarised by J. M. Bowler and R. J. Wasson, 'Glacial Age Environments of Inland Australia', in J. C. Vogel (ed.), *Late Cainozoic Palaeoclimates of the Southern Hemisphere*, A. A. Balkema, Rotterdam, 1984. A trend towards increasing aridity appears to have begun as early as 30,000 years ago with the onset of major aridity from 25,000 to 22,000 years ago. The latter was marked by increasingly mobile dunes in the Simpson Desert and Strzelecki dunefield and by changes in the channel morphology and sediment load of inland river systems such as Coopers Creek and the Finke River.

[47] Ancient wood charcoal and grass phytoliths from archaeological excavations at Puritjarra rock shelter at present provide the only direct evidence of the vegetation of Central Australia between 50,000 and 5,000 years ago. See M. A. Smith, L. Vellen and J. Pask, 'Vegetation History from Archaeological Charcoals in Central Australia: The Late Quaternary Record from Puritjarra Rock Shelter', *Vegetation History and Archaeobotany*, Vol. 4, 1995, pp. 171–177. Also M. A. Smith, 'Environmental History in the Australian Desert: Archaeology, Palaeoenvironments and Changing Landscapes in Western Central Australia', paper presented at the *International Drylands Conference*, International Geological Correlation Program 413, August 2001, Upington, South Africa.

[48] It now seems that people had occupied Central Australia by 30,000 years ago — see M. A. Smith, 'Pleistocene Occupation in Arid Central Australia', *Nature*, Vol. 328, 1987, pp. 710–711; M. A. Smith, B. Fankhauser and M. Jercher, 'The Changing Provenance of Red Ochre at Puritjarra Rock Shelter, Central Australia: Late Pleistocene to Present', *Proceedings of the Prehistoric Society*, Vol. 64, 1998, pp. 275–292; P. B. Thorley, 'Pleistocene Settlement in the Australian Arid Zone: Occupation of an Inland Riverine Landscape in the Central Australian Ranges', *Antiquity*, Vol. 72, 1998, pp. 34–45 — and parts of the Western Desert 24,000 years ago — see S. O'Connor, P. Veth and C. Campbell, 'Serpent's Glen Rockshelter: Report of the First Pleistocene-Aged Occupation Sequence from the Western Desert', *Australian Archaeology*, Vol. 46, 1998, pp. 12–22.

[49] M. A. Smith, 'The Case for a Resident Human Population in the Central Australian Ranges during Full Glacial Aridity', *Archaeology in Oceania*, Vol. 24, 1989, pp. 93–105.

[50] See Mary C. A. Bourke, 'Fluvial Geomorphology and Paleofloods in Arid Central Australia', unpublished PhD thesis, Australian National University, Canberra, 1998.

[51] H. Johnston, P. Clark and J. P. White (eds), *Willandra Lakes: People and Palaeoenvironments, Archaeology in Oceania* (special issue), Vol. 33, 1998.

[52] Jeanette Hope, 'Pleistocene Archaeological Sites in the Central Murray-Darling Basin', in M. A. Smith, M. Spriggs and B. Fankhauser (eds), *Sahul in Review: Pleistocene Archaeology in Australia, New Guinea and Island Melanesia*, Department of Prehistory, Australian National University, Canberra, 1993, pp. 183–196.

[53] Betty Meehan, *Shell Bed to Shell Midden*, Australian Institute of Aboriginal Studies, Canberra, 1982, records one of the best contemporary examples of this, from the Blyth River area of Arnhem Land. During the large-scale flooding of early 1974, the river flowed as freshwater and entirely destroyed the coastal shell-beds that Anbarra people depended upon. It was 'a molluscan eco-disaster', said Meehan. Describing the masses of rotting shellfish, Gurrmanamana, a local man commented 'all the same cemetery'. Although shellfish had been their staple food, the Anbarra switched to alternative resources without significant disruption: wallabies, freshwater turtles, goannas, yams and other vegetables. Many of the latter matured earlier than usual because of the unusual rains.

[54] Dan Gillespie, comment at seminar, Australian National University, 1988.

[55] E. M. Curr, *Recollections of Squatting in Victoria*, Melbourne, 1883 (1965, H. Foster ed.) p. 88, cited in Geoffrey Blainey, *Triumph of the Nomads*, Macmillan, Melbourne, 1975 (first edition), p. 78.

[56] N. B. Tindale, 'Ecology of Primitive Aboriginal Man in Australia', in A. Keast, R. L. Crocker and C. S. Christian (eds), *Biogeography and Ecology in Australia*, Junk, The Hague, 1959, p. 43.

[57] G. Singh and E. Geissler, 'Late Cainozoic History of Vegetation, Fire, Lake Levels and Climate at Lake George, New South Wales, Australia', *Philosophical Transactions of the Royal Society of London*, Vol. B311, 1985, pp. 379–447; A. P. Kershaw, 'An Extended Late Quaternary Vegetation Record from North-Eastern Queensland and its Implications for the Seasonal Tropics of Australia', *Proceedings of the Ecological Society of Australia*, Vol. 13, 1985, pp. 179–189. The sites were Lake George in 1985, and Lynchs Crater in 1986, respectively.

[58] D. Megirian, P. Murray, P. Latz and K. Johnson, 'The Mygoora Lake Local Fauna: A Late Quaternary Vertebrate Assemblage from Central Australia', unpublished manuscript, 2002. For similar published views, see P. Latz, *Bushfires and Bushtucker: Aboriginal Plant Use in Central Australia*, IAD Press, Alice Springs, 1995, pp. 32–37; and A. J. Conacher, 'The Significance of Vegetation, Fire and Man in the Stabilisation of Sand Dunes near the Warburton Ranges, Central Australia', *Earth Science Journal*, Vol. 5, No. 2, 1971, pp. 92–94. All argue that Aboriginal fires led to long-term transformation of the landscape through (re)mobilisation of aeolian landforms. Running counter to these views, J. E. Ash and R. J. Wasson, 'Vegetation and Sand Mobility in the Australian Desert Dunefield', *Zeitschrift für Geomorphologie N. F.* Vol. 45, 1983, pp. 7–25, have argued that the major factor determining dune mobility in Central Australia is wind velocity, not vegetation cover.

[59] D. R. Horton, 'The Burning Question: Aborigines, Fire and Australian Ecosystems', *Mankind*, 1982, pp. 237–251, especially p. 248. The term 'fire-stick farming' was coined by Rhys Jones in 1969, in 'Fire-stick Farming', *Australian Natural History*, Vol. 16, pp. 224–228.

[60] R. L. Clarke, 'Pollen and Charcoal Evidence for the Effects of Aboriginal Burning on the Vegetation of Australia', *Archaeology in Oceania*, Vol. 18, 1983, pp. 35–37.

[61] See papers in R. S. Hill (ed.), *History of the Australian*

Vegetation, Cambridge University Press, 1994. See also A. P. Kershaw, 'Estimates of Regional Climatic Variation within Southeastern Mainland Australia since the Last Glacial Maximum from Pollen Data', *Palaeoclimates*, Vol. 3, Nos 1–3, 1998, pp. 107–134, p. 124: 'the climatic signal from the vegetation is strong ... The general increase in *Eucalyptus* levels through the Holocene may largely be a part of a long-term replacement of fire sensitive by fire promoting vegetation, a process that has been accelerated by the burning activities of Aboriginal people'.

[62] A. P. Kershaw, 'An Extended Late Quaternary Vegetation Record from North-Eastern Queensland and its Implications for the Seasonal Tropics of Australia', *Proceedings of the Ecological Society of Australia*, Vol. 13, 1985, pp. 179–189, especially p. 187.

[63] J. Allen and P. Kershaw, 'The Pleistocene–Holocene Transition in Greater Australia', in L. G. Straus, B. V. Eriksen, J. M. Erlandson and D. R. Yesner, *Humans at the End of the Ice-Age: The Archaeology of the Pleistocene–Holocene Transition*, Plenum, New York, 1996, p. 175.

[64] John Beaton, 'Tasmania — An Extreme Version of Australia', in A. Anderson and T. Murray (eds), *Australian Archaeologist: Collected Papers in Honour of Jim Allen*, Coombs Academic Publishing, Australian National University, Canberra, 2000, p. 184.

[65] J. Allen and P. Kershaw, 'The Pleistocene–Holocene Transition in Greater Australia', op. cit., p. 193.

[66] The phrase 'return of the forests' is from J. Dodson, R. Fullagar and L. Head, 'Dynamics of Environment and People in the Forested Crescents of Temperate Australia', in J. Dodson (ed.), *The Naive Lands: Prehistory and Environmental Change in Australia and the Southwest Pacific*, Longman Cheshire, Melbourne, 1992, p. 113.

[67] The term 'geography of the imagination' is derived from a short story by novelist Graham Greene. In 'The Explorers' (1952) he writes: 'The imagination has its own geography which alters with the centuries', *Collected Essays*, Penguin, 1970, p. 237.

[68] Thomas Carlyle, (19 May 1840), 'A Life. A little gleam of Time between two Eternities,' in 'The Hero as a Man of Letters', in *On Heroes, Hero Worship and the Heroic in History*, Chapman and Hall, London, 1901, p. 176.

[69] Rhys Jones, recorded by Robyn Williams, *The Science Show* (Australian Broadcasting Commission), 2001. Published as 'A Tribute to Rhys Jones', *Australian Archaeology*, Vol. 54, 2002, pp. 59–60. The italics are mine.

Social psychology of climate change

[1] George Marshall, 'Denial and the Psychology of Climate Apathy', *The Ecologist* (UK), November 2001. Marshall draws on the work of Stanley Cohen in *States of Denial*.

[2] George Marshall quotes this from Primo Levi.

[3] *The Australian Financial Review*, 26 March 2001.

[4] Committee on Abrupt Climate Change, *Abrupt Climate Change: Inevitable Surprises*, National Research Council <http://www.nap.edu>.

[5] See, for example, Intergovernmental Panel on *Climate Change, Climate Change 2001: Synthesis Report*, ed. Robert Watson, Cambridge University Press, Cambridge, 2001, p. 343.

[6] In fact, a survey commissioned by the Union of Concerned Scientists in July 2002 concluded that 76 per cent of United States citizens want their government to require power plants and industry to reduce their greenhouse gas emissions. Despite widespread concern, the issue has not gained enough political traction for people to take to the streets over it.

Index